Library of
Davidson College

OUTLINE OF A NOMINALIST THEORY OF PROPOSITIONS

SYNTHESE LIBRARY

STUDIES IN EPISTEMOLOGY,

LOGIC, METHODOLOGY, AND PHILOSOPHY OF SCIENCE

Managing Editor:

JAAKKO HINTIKKA, *Florida State University*

Editors:

ROBERT S. COHEN, *Boston University*

DONALD DAVIDSON, *University of Chicago*

GABRIËL NUCHELMANS, *University of Leyden*

WESLEY C. SALMON, *University of Arizona*

VOLUME 98

PAUL GOCHET

Department of Philosophy, State University of Liège, Belgium

OUTLINE OF A NOMINALIST THEORY OF PROPOSITIONS

*An Essay in the Theory of Meaning
and in the Philosophy of Logic*

D. REIDEL PUBLISHING COMPANY

DORDRECHT : HOLLAND / BOSTON : U.S.A.
LONDON / ENGLAND

Library of Congress Cataloging in Publication Data

Gochet, Paul.
　　Outline of a nominalist theory of propositions.
　　(Synthese library ; v. 98)
　　Revised version (and translation) of Esquisse d'une théorie nomina-
liste de la proposition, published by Armand Colin, Paris, 1972.
　　Includes bibliographical references and indexes.
　　1. Proposition (logic). I. Title.
BC181.G5613　　160　　80–13056
ISBN 90–277–1031–7

Published by D. Reidel Publishing Company,
P.O. Box 17, 3300 AA Dordrecht, Holland

Sold and distributed in the U.S.A. and Canada
by Kluwer Boston Inc., Lincoln Building,
160 Old Derby Street, Hingham, MA 02043, U.S.A.

In all other countries, sold and distributed
by Kluwer Academic Publishers Group,
P.O. Box 322, 3300 AH Dordrecht, Holland

D. Reidel Publishing Company is a member of the Kluwer Group

All Rights Reserved
Copyright © 1980 by D. Reidel Publishing Company, Dordrecht, Holland
No part of the material protected by this copyright notice may be reproduced or
utilized in any form by or any means, electronic or mechanical,
including photocopying, recording or by any informational storage and
retrieval system, without written permission from the copyright owner

Printed in The Netherlands

To the Memory of Philippe Devaux

ACKNOWLEDGMENTS

This book is an extensively revised version of *Esquisse d'une Theorie Nominaliste de la Proposition* published by Armand Colin, Paris, 1972.

The initial translation of the French version cited above was made by Margret Jackson. The author, who has carried out an extensive revision of the French text, wishes to express his gratitude to both Margret and Howard Jackson and also to acknowledge the invaluable assistance of Anthony Dale, Susan Stern-Gillet and Alan S. Beedle who assisted him in the translation of the revised sections.

Many important changes in the content of this revised edition are due to criticisms and suggestions made by the Reidel referee and also by Anthony Dale. Michael More accomplished the unrewarding task of reading the proofs.

Needless to say, the above named persons are not to be held responsible for any errors which may remain in the text.

TABLE OF CONTENTS

ACKNOWLEDGMENTS   vii

INTRODUCTION   1
1. Importance of the Subject   1
2. The Roles Played by the Concept of Proposition   1
3. How to Conceive of a Theory of Proposition   3
4. Which Method to Use   3
5. The Merits of Nominalism   5
6. Varieties of Nominalism   7
7. The Senses of the Word 'Proposition'   11

CHAPTER I / THE CRITERION OF ONTOLOGICAL COMMITMENT   15
1. Quine's Criterion of Ontological Commitment   15
2. Warnock's Objections to Quine's Criterion of Ontological Commitment   18
3. The Application of the Criterion of Ontological Commitment to Propositions   22
4. Compromising Uses of the Word 'Proposition'   24
5. Critique of Ayer's First Attempt to Escape Ontological Commitments to Propositions   25
6. The Double Interpretation of the Existential Quantifier   26
7. The Double Interpretation of Bound Variables   28
8. From Pragmatics to Ontology   31

CHAPTER II / THE SYNTACTIC APPROACH   34
1. Is an Axiomatic Definition of Proposition Possible?   34
2. Two Nominalist Solutions on the Problem of Interpreting Propositional Variables   35
3. What Quine's Notation Reveals With Regard to the Status of Propositions   37
4. Does the Definition of Logical Truth Presuppose the Concept of Proposition? Strawson's Thesis   40
5. Replies to Strawson's Objections   41

6. The Definition of Proposition in Terms of the Premises and Conclusion of an Inference 43

CHAPTER III / A SEMANTIC DEFINITION OF PROPOSITION IN TERMS OF TRUTH AND FALSITY 45
1. The Aristotelian Definition of Proposition in Terms of Truth 45
2. The Influence of the Semantic Definition of Truth on the Concept of Proposition 46
3. Use of the Distinction Between Sentence and Statement as a Solution to the Paradox of the Liar in Natural Language 49
4. The Ontological Status of the Distinction between Statements and Sentences 51
5. Truth and Falsity Apply to Sentences Before Applying to Statements 54
6. The Semantic Theory of Truth and the Correspondence Between Language and Reality 55

CHAPTER IV / THE PRAGMATIC DEFINITION OF PROPOSITION IN TERMS OF ASSERTION OR ASSERTABILITY 60
1. The Pragmatic Definition of Proposition in Terms of Assertability 60
2. The Distinction Between Proposition and Statement from a Pragmatic Perspective 62
3. Austin's Distinction Between Locutionary and Illocutionary Acts 63
4. An Examination of Searle's Notion of Proposition 64
5. Stenius' Analysis 66
6. The Performative Hypothesis 67
7. Hausser's Treatment of Moods 68
8. A Vindication of Searle's Position 69
9. A New Account of Searle's Concept of Propositional Content 71

CHAPTER V / THE NATURE OF FACTS 73
1. The Nature and Status of Facts in Russell's 'Philosophy of Logical Atomism' 73
2. The Merits of Russell's Notion of Fact 74
3. The Defects of Russell's Theory of Facts 76
4. Wittgenstein's Conception of Fact 77
5. Arguments For and Against the Ontological Interpretation of Facts 79
6. Application of Methods of Generative Grammar to Detect the Ontological Nature of Facts 82
7. Why There Cannot be Facts 83

CHAPTER VI / THE PROPOSITION IN TERMS OF BELIEF

1. Belief and Proposition — 87
2. The Problem of False Beliefs — 89
3. The Distinction Between Propositional Verbs and Cognitive Verbs — 92
4. The Logical Syntax of Propositional Verbs — 93
5. An Attempt at Absorbing Propositions into Sentences — 96
6. Searle's Views on Intentionality — 98

CHAPTER VII / PROPOSITIONS AS MEANINGS OF SENTENCES — 101

1. The Relational Conception of Meaning — 101
2. The Eternality and Temporality of Meaning — 102
3. The Behaviouristic Analysis of the Meaning of Sentences — 104
4. The Chess-Theory of Meaning — 106
5. An Attempt at Dissolving the Problem Raised by the Meaning of Sentences — 108
6. The Picture Theory of Meaning — 109
7. The Limitations of the Picture Theory of Meaning — 111
8. Beyond the Picture Theory — 113
9. The Recursive Definition of Truth as a Tool for Compositional Semantics — 116
10. Recursive Semantics and Nominalism — 117
11. Categorial Grammar, Set Theoretic Semantics and Nominalism — 120
12. Game-Theoretical Semantics — 120

CHAPTER VIII / AN ATTEMPT AT A NEW SOLUTION FOR THE ENIGMA OF THE MEANING OF FALSE SENTENCES — 124

1. Conditions of Adequacy on a Satisfactory Answer — 124
2. Ryle's Solution to the Enigma of the Meaning of False Sentences — 125
3. The Possibility of Falsity as a By-Product of the Creativity of Language — 126
4. The Solution Offered by Possible Worlds Semantics to the Enigma of the Meaning of False Sentences — 127
5. A Pragmatic Solution of the Enigma — 128
6. Nominalism Again — 129

CHAPTER IX / THE IDENTIFICATION CRITERION OF PROPOSITIONS — 131

1. The Importance of Finding a Criterion of Propositional Identity — 131
2. The Definition of Proposition in Terms of Synonymy — 132

3. Intensional Isomorphism	134
4. The Role of the Notion of Isomorphism in Defining a Criterion for the Identity of Propositions	135
5. Preliminaries to the Application of the Criterion of Extensional Isomorphism	138
6. Some Final Refinements of the Notion of Extensional Isomorphism	142
7. Vanderveken's Criterion	142
8. Suppes' Gradualism	144
9. Indeterminacy of Translation	145
CHAPTER X / PROPOSITIONS AND INDIRECT DISCOURSE	149
1. The Notion of Proposition and of Indirect Discourse	149
2. The Syntactic Approach to the Problem of Intensional Contexts	150
3. Prior's Nominalist Syntax	154
4. L.J. Cohen's Extensionalist Syntax	156
5. Frege's Dualist Semantics and Epistemic Logic	158
6. Carnap's Dualist Semantics	159
7. Quine's Unitary Extensionalism	161
8. Criticisms Addressed to Quine's Nominalist Theory: Kaplan's Alternative Solution	166
9. Hintikka's Pluralistic Extensionalism	169
10. The Pragmatic Approach to the Problem of Intensional Contexts: Natural Pragmatics	174
11. The Pragmatic Approach to the Problem of Intensional Contexts: Formal Pragmatics	178
12. Objections Against Montague's Semantics	181
CONCLUSION	185
NAME INDEX	193
SUBJECT INDEX	196

INTRODUCTION

1. IMPORTANCE OF THE SUBJECT

In 1900, in *A Critical Exposition of the Philosophy of Leibniz*, Russell made the following assertion: "That all sound philosophy should begin with an analysis of propositions is a truth too evident, perhaps, to demand a proof".[1] Forty years later, the interest aroused by this notion had not decreased. C. J. Ducasse wrote in the *Journal of Philosophy*: "There is perhaps no question more basic for the theory of knowledge than that of the nature of propositions and their relations to judgments, sentences, facts and inferences".[2] Today, the great number of publications on the subject is proof that it is still of interest. One of the problems raised by propositions, the problem of determining whether propositions, statements or sentences are the primary bearers of truth and falsity, is even in the eyes of Bar-Hillel, "one of the major items that the future philosophy of language will have to discuss".[3]

Ph. Devaux gave a correct summary of the situation when he wrote in his *Russell* (1967):

Since Peano and Schröder who, in fact, adhered more faithfully to Boole's logic of classes, the logical and epistemological status of the proposition together with its analysis have not ceased to be the object of productive philosophical controversies. And especially so since the establishment of contemporary symbolic logic, the foundations of which have been laid out by Russell and Whitehead.[4] *

2. THE ROLES PLAYED BY THE CONCEPT OF PROPOSITION

The concept of proposition is central to *logic*, and it even serves to define this discipline. Strawson writes: "If we use 'proposition' as a general name for what, when these forms are exemplified, we introduce or specify by such 'that'-, 'whether'- or 'if'-clauses, then logic is the general theory of the proposition".[5] The concept of proposition holds a privileged position in logic, but Strawson goes too far with his definition of logic. His definition turns into a contradiction a logic of questions or a logic of imperatives in

---

* *Editorial Note*: All quotations originally written in languages other than English appear in translation herein.

which questions or imperative sentences would not be analyzable in terms of propositions. This seems arbitrary.

The *theory of knowledge* is equally in need of this concept in order to define its object or, at least, one of its objects, since, as Russell has said, there are at least two ways of understanding the word 'know':

> In ...[one] use it is applicable to the sort of knowledge which is opposed to error, the sense in which what we know is *true*, the sense which applies to our beliefs and convictions, i.e. to what are called *judgements*... . In [a] second use..., the word applies to our knowledge of *things,* which we may call *acquaintance.*[6]

Propositions have been a subject of *ontological* controversy ever since philosophers attributed to them the status of independent entities. Bolzano understood them in this way in his *Wissenschaftslehre* of 1837 (where he speaks of *Satz an sich*), and so did Moore in 'The Nature of Judgement' (1899), an article one might take as the manifesto of British neo-realism. In this paper, believing that it is necessary to reify meaning in order to guarantee the independence of truth in relation to the thinking subject, Moore develops an obviously Platonistic conception of propositions. He writes: "Existence is itself a concept; ... truth cannot be defined by a reference to existence, but existence only by a reference to truth".[7]

Within the framework of *philosophical psychology* propositions are a main element of the theory of intentionality, according to which mental acts (belief, judgment, fear, love) would all be directed at a single object. Although in the case of knowledge facts may serve the role of objects, in the case of belief (particularly false belief), this solution is excluded and one will have to have recourse to propositions in place of facts.

Finally, propositions are invoked to account for the meaning of sentences in a *theory of meaning*. When one translates a text from one language into another, the linguistic form is altered; but the content, it is claimed, remains unchanged. Propositions are identified with this invariant that survives translation.

Philosophers who have tried to define the notion of proposition and to determine their ontological status have generally dealt with only one of the roles of that notion. Numerous ramifications into all domains of philosophy and the intricacy of problems raised by the concept of a proposition suggest that a comprehensive treatment is urgently needed. One may foresee that this will be a difficult task. This degree of complexity, however, is not unusual in philosophical analysis. If the latter is intricate, it is as Montefiore writes, "because it nearly always involves keeping in mind a great number of points at once, and the ability to follow a sustained and systematic argument".[8]

3. HOW TO CONCEIVE OF A THEORY OF PROPOSITIONS

We take the term 'theory' in the sense given it by Castañeda:

> Now, a theory (whether philosophical or scientific) is *not* the sort of thing that comes out as the conclusion of a proof.... . A philosophical theory is a point of view deployed in a purported systematic, harmonious organization of initially perplexing philosophical facts in such a way that the initial perplexities vanish. But a theory does allow of conclusive refutation by being shown that it is not capable of covering certain facts, i.e., of "saving certain appearances", as the Greek philosophers used to say.[9]

The internal *cohesion* of a theory does not, therefore, simply consist in *consistency*, that is, in the absence of contradiction. In constructing a theory, it is not sufficient to juxtapose fragmentary solutions and eliminate those which turn out to be incompatible, as the proponents of eclecticism might argue.

A theory does not merit the name unless it is capable of solving by itself, i.e. *without external contribution of ad hoc solutions*, many problems it was not especially designed to solve. For this reason, the best way of *confirming* a theory should be to actively elicit problems that are within its reach and to see how it deals with them. This has already been affirmed by Russell in *On Denoting* (1905):

> A logical theory may be tested by its capacity for dealing with puzzles, and it is a wholesome plan, in thinking about logic, to stock the mind with as many puzzles as possible, since these serve much the same purpose as is served by experiments in physical science.[10]

In this spirit, I will show that the *same* conception of proposition may provide an answer to questions as *different* as the following three:

(1) How can sentences with the *same* truth − value be true of *different* things?

(2) How can *false* sentences have a meaning?

(3) How can we *immediately* give a meaning to sentences never heard before (problem of the creativity of language)?

4. WHICH METHOD TO USE

The problems to be discussed are all more or less closely linked to a traditional philosophical problem: the problem of universals. Today it has to be approached in a new way; for new methodological exigencies have been imposed upon the philosophers dealing with it, as W. Stegmüller[10] points out very clearly in *Main Currents in Contemporary German, British and American Philosophy* (1969). Since our research has to be guided by these demands, it is worthwhile to recall what they consist of:

We can no longer be satisfied with advancing *a priori* arguments for or against one or another view. There is a further question that we must consider above all: whether a given viewpoint on this matter is compatible with the preservation of the overall content of contemporary science, and does not destroy it in whole or in part. *The fact that the anti-Platonist viewpoint cannot be refuted does not suffice to establish its acceptability* – at least, not if we grant the requirement *that a solution to the problem of universals, however constituted, must not lead to such an impoverishment of our system of concepts and judgments that basic sciences are perforce not merely reformulated but abandoned.*[11]

The desire to conform to these legitimate demands will, for example, cause us to reject as unacceptable certain extreme forms of nominalism which would oblige us to renounce *set theory*. The question becomes more delicate as soon as there is a conflict between a certain solution of the problem of universals and a science which is not a basic one. A conflict of this kind opposes the doctrine of extensionalism – i.e., the recognition of a principle of exchange which allows the interchangeability of coreferential expressions, of coextensive predicates and of materially equivalent sentences, – with the existence of apparent counter-examples which seem to support intensional logic. Are not the scruples of the extensionalist, and *a fortiori* those of nominalists, conducive to a 'Malthusian' attitude harmful to science? We still have to pose this question.

The role that I am going to attribute to science is even greater than Stegmüller attributed to it. In effect, I will have to invoke scientific theories not so much to *overturn* philosophical positions, but rather, in order to *establish* new ones. I will reflect less about the naked or day-to-day reality but, rather, about "reality as shown by science", to use the very appropriate term used to describe the program of the journal *L'Age de la Science*. The sciences to which I will constantly have to refer are symbolic logic, linguistics, and to a lesser extent, psychology.

Here I will differ from the analysts of natural language (Strawson, Hampshire), as well as from the theoreticians of formalized languages (Tarski, Carnap). None of them, incidentally, found it necessary to consult with linguists, an omission for which the former have been criticized by H. Hubien[12] in 'Philosophie analytique et Linguistique moderne' (1968), the latter by Bar-Hillel[13] in 'Do Natural Languages Contain Paradoxes?' (1966). The fact that linguistics did not have anything to offer philosophy at that time, however, may, as Hubien correctly observes, very well account for this indifferent attitude toward scientific linguistics, for which they are reproached today.

One of the most clairvoyant among these philosophers did foresee the establishment of a real science of language. In *Ifs and Cans* (1956), one year before Chomsky inaugurated transformational linguistics, Austin wrote these prophetic words: "Is it not possible that the next century may see the birth, through the joint labours of philosophers, grammarians, and numerous other students of language, of a true and comprehensive *science of language*?"[14]

5. THE MERITS OF NOMINALISM

The nominalist spirit which will inspire this research requires some justification all the more so as the term can cover very different programs.

Nominalism — as I conceive it — does not come from a bias against abstract entities comparable with the empiricist bias against unobservable entities. I by no means eschew all abstractions. Those I find objectionable are those which lack explanatory power, those which are accepted *praeter necessitatem*.

The kind of Nominalism I subscribe to from the start could be labelled *methodical nominalism* as opposed to *doctrinal nominalism*. I adhere to Russell's basic principle of Methodical Nominalism, i.e. to a law of economy which Vuillemin characterized in these terms in his *Leçons sur la première philosophie de Russell* (1968):

> Whatever can be logically constructed — starting with logical notions — is not Real. If, in a system of entities thought to be primitive, analysis shows that some of these entities can be logically constructed out of others, they must be eliminated from the inventory of reality. Reality thus consists of what resists logical analysis conducted according to the principle of parsimony.
> Such is the objective role of the law. It is the inversely proportional measure of Reality.[15]

This principle attracted many criticisms. For instance, it could be argued against Vuillemin that Russell's logical constructions do not dispense with the constructed objects. Some of these criticisms arise from a misunderstanding. Let me clear up one of them right away.

In 'Names and Descriptions' (1962), relying on the thesis that proper names can be eliminated from the *language*, A. J. Ayer infers that the ontological category of substance can be dispensed with. Such an inference invokes the law of economy. Professor Dopp, however, claims that Ayer is committed to something more, to some variety of ontologico-grammatical parallelism, an outdated doctrine already discarded by Aristotle's theory of abstraction:

> ... we wonder how an analysis of language... could shed light on the ontological nature of the ultimate constituents of reality.

> Why should distinct categories of words in the language (or 'ideas' of human thought) necessarily be correlated with the distinct categories of elements 'constituting' the Real? It is precisely this [sort of relation] which the Aristotelian theory of abstraction was understood to deny.[16]

Dopp's objection to Ayer's claim seems to be rooted in the failure to recognize an important asymmetry between two kinds of philosophies. In the philosophies of the former kind, ontological conclusions are drawn from the fact that we *need* signs of a certain sort. In those of the latter, ontological conclusions are derived from the fact that we *can* do *without* those signs. Dopp addresses the same objections to both. It is my claim, that only the former type of inference is objectionable. As Ayer appeals to the latter, but not to the former, his inference is immune to Dopp's criticism.

The *need* for signs of a certain sort is not enough to establish the *existence* of ontological entities correlated with them. Such a need might indeed result from a syntactic constraint intrinsic to the language under consideration. On the contrary, only the *dispensability* of signs of a certain category allows ontological inferences. Admittedly, it does not prove that the entities correlated with the signs dispensed with do not exist, but it shifts the burden of proof. It is up to the philosopher who advocates the existence of *substances* or *propositions* to argue for his tenet. The dispensability of the sign is, *prima facie*, a good reason for the dispensability of the *significatum* or at least it removes one reason to posit such entities.

Now that Dopp's objection has been cleared away, I have to offer positive justification for the law of economy. It will be enough here to refer to Wittgenstein's argument in the *Tractatus* (as rightly invoked by Vuillemin in this connection):

> 5.47321 Occam's maxim is, of course, not an arbitrary rule, nor one that is justified by its success in practice: its point is that *unnecessary* units in a sign-language mean nothing.
>
> Signs that serve *one* purpose are logically equivalent, and signs that serve *none* are logically meaningless.[17]

The kind of nominalism I began with is *methodical nominalism*. The kind of nominalism I will come to is *doctrinal nominalism*. The former is a *rule* that is rationally grounded upon methodological considerations such as the burden of the proof or the power of an explanation. The latter is a *theory* which tries to replace *metaphysical* entities by *linguistic* entities whenever possible or to make the postulation of platonic entities appear unnecessary thanks to some sort of paraphrase. *The distinction between these two kinds of nominalism* is worth stressing. I would invoke that very distinction were I being accused of begging the question in favour of nominalism.

INTRODUCTION   7

It should be noticed that the concept of *nominalism* should be carefully dissociated from that of *reductionism*. Nominalistic ascent is not necessarily a kind of reduction as it would be if language were simple. But it is not. "Everyday language is a part of the human organism and is no less complicated than it".[18]

Austin's and Chomsky's discoveries support Wittgenstein's claim. Austin has shown that certain actions, such as promising, bequeathing and so forth, are part and parcel of language understood as a social institution. A new terminology had to be forged in order to describe these sophisticated uses of language where 'we do things with words'. Neither physics nor psychology offered concepts suitable for describing these linguistic facts.

Chomsky also reached the conclusion that linguistic competence cannot be accounted for in terms of habits or dispositions and appeals to analogy which would be common to linguistic and non-linguistic kinds of competence. He stated that "it is quite impossible to provide an explanation in terms of 'habits' and 'dispositions' and 'analogy' ";[19] and later added, "There seems to be no substance to the view that human language is simply a more complex instance of something to be found elsewhere in the animal world".[20]

In the same spirit, I will try to establish that the solution of a problem such as the meaningfulness of a false sentence cannot be solved unless one is prepared to take account of language conceived as a *sui generis* phenomenon. Any attempt at reducing language to a simpler sort of phenomena is doomed to failure.

6. VARIETIES OF NOMINALISM

There are several kinds of doctrinal nominalism. The kind which will be defended here would be more correctly called 'Extensionalism'. To appreciate it correctly one has to set it in contrast both to Goodman's nominalism and to Carnap's and Church's intensionalism.

(A) *Goodman's Nominalism*

"Nominalism as I conceive it", Goodman writes in 'A World of Individuals' (1956) "...does not involve excluding abstract entities, spirits, intimations of immortality, or anything of the sort; but requires only that whatever is admitted as an entity at all be construed as an individual".[21] Goodman thus rejects classes and properties but also propositions.

How does Goodman's nominalism differ from Quine's extensionalism? Goodman accepts in his ontology individuals and aggregates of individuals, but he rejects classes. The difference between the two is this: whereas two aggregates only differ when some entity among those which make the aggregates belongs to one of them only, two classes can differ even though they share all their *elements*. In other words, *class membership* suffices to generate new classes while an element must be added or withdrawn to generate new aggregates. The aggregate [[a, b], [c, d]] is the same as [[a, c] [b, d]], whereas the class {{a, b}, {c, d}} is not the same as {{a, c}, {b, d}}. Or, to borrow an example from Lesnieuwski's mereology which anticipated Goodman's calculus of individuals, "a sphere is identical with the class$_L$ [class in the sense of a whole] of its halves and the class$_L$ of its quarters".[22]

The reason why Goodman rejects classes and adopts aggregates is this: classes one recognizes expand indefinitely. Aggregates, on the other hand, cannot be generated *ad libitum* out of their components: for n components there are only $2^n - 1$ aggregates.

(B) *Extensionalism*

Quine's extensionalism is less drastic than Goodman's nominalism. Quine tolerates classes in his *ontology* but rejects concepts or propositions. He acknowledges the need for the extension—intension dichotomy, but he refuses to match this semantical distinction (which goes back to Aristotle) with an ontological dichotomy. Though he is prepared to assume the existence of classes, he refuses to commit himself to that of concepts or propositions.

The reason behind Quine's scruple is this: we possess a criterion of identity for classes but we lack such a criterion for concepts or propositions.[23] Two classes are identical whenever they share all their members. In short, classes obey the principle of extensionality which reads as follows

$$(\alpha = \beta) \equiv (x)(x \in \alpha \cdot \equiv \cdot x \in \beta)$$

The class of human beings is identical with that of featherless bipeds since both classes have the same members. The same holds for the class of the passengers on the Mayflower and the founders of the Plymouth colony.

Concepts do not obey that principle. The concept of humanity does not coincide with the concept of featherless biped even if all human beings *are* featherless bipeds and only they. To grasp this point, think of a world where all round objects are red (and vice-versa). It would not follow that the concept of redness was identical with the concept of roundness.

Should we say that two concepts differ as soon as they are not analytically equivalent? Quine has objected that since the open sentence 'Fx & p' where 'p' is contingently true would not be analytically equivalent to 'Fx', we could in this way *indefinitely* and — what is worse — *arbitrarily* multiply intensional entities such as concepts and propositions.[24]

Extensionalism cuts across this inflation of concepts by ascribing one and the same extension to all coextensive predicates. It can therefore be described as a *mitigated form* of nominalism and conversely extensionalism can be said to be a first step towards strict nominalism. Both doctrines spring from a common aversion to an unnecessary multiplication of entities and differ only in degree. As Goodman observes: "Extensionalism precludes the composition of more than one entity out of exactly the same entities by membership; nominalism goes further, precluding the composition of more than one entity out of the same entities by any chains of membership".[25] For a strict nominalist like Goodman — but not for Quine — second-order membership *collapses* with first-order membership and the aggregate of the aggregate coincides with the initial aggregate.

Notice, however, how much Quine's extensionalism differs from traditional nominalism. Far from reducing universals to words, Quine acknowledges that the number of universals exceeds the number of constructible names:

...through our variables of quantification we are quite capable of committing ourselves to entities which cannot be named individually at all in the resources of our language; witness the real numbers, which, according to classical theory, constitute a larger infinity than does the totality of constructible names in any language.[26]

Quine's ontology thus accounts for the uncountable to which Cantor gave access. Such a recognition of an excess of sets over names should not be interpreted as an adherence of Quine to some sort of Platonistic metaphysics. What convinces Quine of the need to accommodate transfinite sets is not an argument like Plato's dialectics but a scientific argument: Cantor's diagonal.

(C) *Intensionalism*

In *Meaning and Necessity* (1947),[27] Carnap developed a theory of universals which can rightly be called intensionalism. It has just been said that intensional entities, such as concepts and propositions, do not obey the extensionality principle: i.e. that for them the principle where 'f' is a predicate expression,

$$(\Phi)(\Psi)\,[(\Phi = \Psi) \equiv (f\Phi \equiv f\Psi)],$$

is not valid.

Carnap proposed that a new principle, which one might dub the 'principle of intensionality', should act as proxy for the classical principle of extensionality. This new principle reads as follows: Two concepts are identical if and only if all the objects which exemplify the former exemplify the latter, and conversely. This can be formally expressed as

$$\Phi = \Psi \equiv (x)(\Phi x \leftrightarrow \Psi x),$$

or

$$\Phi = \Psi \equiv \Box (x)(\Phi x \equiv \Psi x),$$

or again

$$\Phi = \Psi \equiv \{ `(x)(\Phi x \equiv \Psi x)\text{' is L-true}\},$$

where L-true means, as soon as meaning postulates are incorporated 'analytically true'.

For example, the concept of 'bachelor' is identical to the concept of 'unmarried man' because 'x is a bachelor' is tied- up with 'x is an unmarried man' by a relationship of strict bi-implication in contradistinction to 'x is a passenger on the Mayflower' with respect to 'x is a founder of Plymouth'. *Mutatis mutandis*, two sentences express the same proposition when they are L-equivalent and (Carnap adds) intensionally isomorphic.

As applied to Carnap's doctrine, the label 'intensional Platonism' might be inappropriate. Not so, however, when it is used in connection with Church. The latter advocated a frankly realistic account of proposition in his contribution to the *Encyclopedia Britannica* under the entry 'Proposition':

For some purposes at least there is needed a more abstract notion, independent alike of any particular expression in words and of any particular psychological act of

ONTOLOGICAL POINTS OF VIEW AND KINDS OF LANGUAGES:

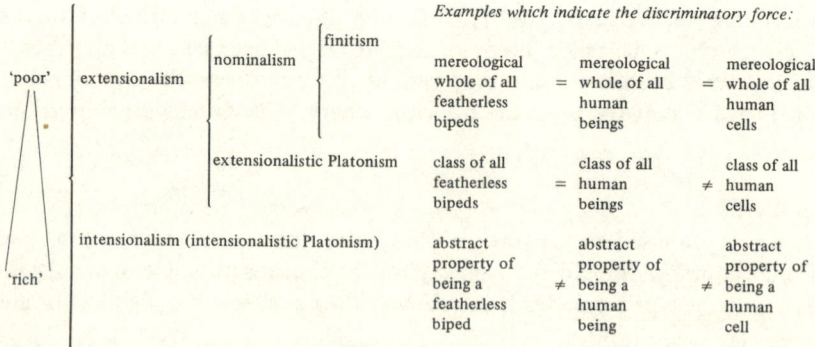

judgment or conception – not the particular declarative sentence, but the content of meaning which is common to the sentence and its translation into another language – not the particular judgment, but the objective content of the judgment which is capable of being the common property of many.[28]

The three main positions with respect to the problem of universals have been summarized by Küng. I borrow the following map (shown opposite) from his *Ontology and the Logistic Analysis of Language* (1967).[29]

7. THE SENSES OF THE WORD 'PROPOSITION'

I have stated how the word 'theory' as it occurs in this essay, ought to be understood. Also the word 'Nominalism' has been defined. It remains to examine the meanings associated with the word 'proposition'.

The definitions presented so far in the literature vary with the context set up by the particular problem propositions raise. I will go through each of them and try to assess their ontological implications.

The following table will guide this enquiry:

1. *Logic*

    1.1. A proposition is whatever can be a premise or a conclusion of an argument or what can be in "– in the contradictory of..." relation. It is also the class of interdeducible sentences. (*Syntax*)

    1.2. A proposition is what 'true' or 'false' can be predicated of. (*Semantics*)

    1.3. A proposition is what can be asserted or hypothesized. (*Pragmatics*)

2. *Ontology*

    2.1. A proposition identifies a fact.

    2.2. A true proposition is a fact.

    2.3. A proposition is a possible state of affairs.

3. *Philosophy of mind*

    3.1. A proposition is the second term of the dyadic relation expressed by the verb 'to believe'.

    3.2. A proposition is what is thought.

4. *Theory of meaning*

    4.1. A proposition is the second term of the dyadic relation expressed by the verb 'to mean' when the first term denotes a sentence.

    4.2. A proposition is the class of all sentences synonymous with a well-formed sentence in a given language.

4.3. A proposition is the class of all sentences standing in the relationship of extensional isomorphism with one another.

4.4. A proposition is the class of all sentences standing in the relationship of intensional isomorphism with one another.

This table presents more than a survey. It will be used as a frame to guide the theoretical work to be accomplished. This confers a precise content to the requirement of *systematicity* which was formulated before. The notion of proposition will be given a definition or analysis which accounts for the multifarious roles attributed to it in the table.

This effort towards unification is not gratuitous. It is needed for two reasons: on the one hand, because partial definitions of the notion of proposition must fit together; and, on the other hand, because the same concepts occur in different definitions. I claim for these recurring concepts the same sort of invariance as Davidson requires of the concept of the 'logical form of a sentence' which occurs in several contexts. Observing that the notion of 'logical form of a sentence' occurs both in the theory of truth and in the theory of logical consequence, Davidson rightly says that "the two approaches to logical form cannot yield wholly independent results...".[30]

Vendler's paper on 'Les Performatifs en perspective' (1970), could serve as a counter-proof to demonstrate the perils one must face when tackling problems in isolation. Approaching propositions from the viewpoint of the analysis of language, Vendler writes that "...it is clear, and this is contrary to prevalent usage, that propositions, in themselves, are not true or false,... Only propositions issued with a certain force can be true or false...".[31] Now, as Geach pointed out in 'Assertion' (1965), such a view, taken literally, would destroy so fundamental a part of logic as the theory of truth-functions.

If this [that a sentence can have a truth value assigned to it only in that it is 'used to make a statement' in a given context] were literally true, then a truth-functional account of '*p vel q*' or of '*p aut q*' would be impossible: for the disjunct clauses represented by '*p*' and '*q*' would not be being 'used to make statements' in a context in which only the disjunction was asserted, and would thus not have any truth values for the truth value of the whole proposition to be a function of.[32]

In accordance with Stegmüller's methodological principles to the effect that a satisfactory solution to the problem of universals must not lead to a mutilation of science, we ought to reject Vendler's view on propositions not being truth-bearers.

To substantiate the claim that one can dispense with propositions, it is not enough to show that one can dispense with the word 'proposition'. On the other hand, some uses of the word 'proposition' are not ontologically loaded. When is one *committed* to propositions? The first chapter[33] will be devoted to this preliminary question.

REFERENCES

[1] B. Russell, *A Critical Exposition of the Philosophy of Leibniz*, Cambridge University Press, 1900, p. 8.
[2] C.-J. Ducasse, 'Propositions, Opinions, Sentences and Facts', *Journal of Philosophy* 37 (1940) 701.
[3] Y. Bar-Hillel, 'Universal Semantics and Philosophy of Language', in J. Puhvel (ed.), *Substance and Structure of Language*, University of California Press, Berkeley and Los Angeles, 1969, pp. 16–17.
[4] Ph. Devaux, *Bertrand Russell ou la Paix dans la vérité*, Seghers, Paris, 1967, p. 59.
[5] P. F. Strawson, 'Introduction', in P. F. Strawson (ed.), *Philosophical Logic*, Oxford University Press, 1967, p. 1.
[6] B. Russell, *The Problems of Philosophy*, Oxford University Press, 1952, p. 44.
[7] G. E. Moore, 'The Nature of Judgment', *Mind* 8 (1899) 180.
[8] A. Montefiore, *A Modern Introduction to Moral Philosophy*, Routledge and Kegan Paul, London, 1958, p. 1.
[9] H. N. Castañeda, 'Ethics and Logic: Stevensonian Emotivism Revisited', *Journal of Philosophy* 64, (1967) 673–674.
[10] B. Russell, 'On Denoting', reprinted in H. Feigl and W. Sellars (eds.), *Readings in Philosophical Analysis*, Appelton-Century-Crofts, New York, 1949, p. 108.
[11] W. Stegmüller, *Main Currents in Contemporary German, British and American Philosophy*, D. Reidel, Dordrecht, 1969, p. 54.
[12] H. Hubien, 'Philosophie analytique et Linguistique moderne', *Dialectica* 22, (1968) 96.
[13] Y. Bar-Hillel, 'Do Natural Languages Contain Paradoxes?', *Studium generale*, (1966) 393. Reprinted in *Aspects of Language*, North Holland, Amsterdam 1970; see p. 276.
[14] J. L. Austin, 'Ifs and Cans', *Proceedings of the British Academy*, Vol. XLII, Oxford University Press, London, 1956, pp. 131–132. Reprinted in J. L. Austin, *Philosophical Papers*, Oxford University Press, 2nd. edn., 1970; see p. 232.
[15] J. Vuillemin, *Leçons sur la première philosophie de Russell*, A. Colin, Paris, 1966, p. 266.
[16] J. Dopp, 'Thinking and Meaning', Entretiens d l'Institut International de Philosophie, 1962, *Logique et Analyse* 20, (1962) 207.
[17] Wittgenstein, *Tractatus Logico-philosophicus*, 5.47321, with a new translation by D. F. Pears and B. F. McGuinnes and Introduction by B. Russell, Routledge and Kegan Paul, London, 1961, p. 97.
[18] Wittgenstein, *Ibid.*, 4.002, p. 35.
[19] N. Chomsky, *Language and Mind*, Harcourt Brace, 1968; enlarged edn., 1972, p. 48.
[20] Chomsky, *Ibid.*, p. 70.
[21] N. Goodman, 'A World of Individuals' in I. M. Bochenski, A. Church, N. Goodman, *The Problem of Universals*, Notre Dame University Press, 1956, p. 12.

[22] G. Küng, *Ontology and the Logistic Analysis of Language*, (trans. by E.C.U. Mays), D. Reidel, Dordrecht, 1967, p. 106.
[23] W. V. O. Quine, *Word and Object*, M.I.T. and Wiley, 1960, p. 209.
[24] W. V. O. Quine, *From a Logical Point of View*, Harper Torchbooks, New York, 1961, p. 153.
[25] Goodman, *Op. cit.*, p. 19.
[26] W. V. O. Quine, 'Carnap's Views on Ontology' (1951); repr. in *The Ways of Paradox and other Essays*, Random House, New York, 1966, p. 128.
[27] R. Carnap, *Meaning and Necessity* (1947) University of Chicago Press, enlarged edn., 1956, § 5 and § 33.
[28] A. Church, 'Proposition', *Encyclopedia Britannica*, U.S.A., Vol. 18, 1965, p. 560.
[29] Küng, *Op. cit.*, p. 139.
[30] D. Davidson, 'On Saying That' *Synthese* 19, (1968) 132.
[31] Z. Vendler, 'Les Performatifs en perspective', *Langage* 17, (1970) 89–90.
[32] P. T. Geach, 'Assertion', *Philosophical Review* 74, (1965) 452–453.
[33] For a study of the proposition in ancient and medieval philosophy, the reader is referred to G. Nuchelmans' important monograph *Theories of Proposition. Ancient and Medieval Conceptions of the Bearers of Truth and Falsity*, North Holland, Amsterdam, 1973.

CHAPTER I

THE CRITERION OF ONTOLOGICAL COMMITMENT

1. QUINE'S CRITERION OF ONTOLOGICAL COMMITMENT

In 'Ontological Commitment' (1958), A. Church maintains that a *precise* formulation of a criterion of ontological commitment is a *prior condition* of any fruitful and rigorous discussion of the problem of universals: "... no discussion of an ontological question, in particular of the issue between nominalism and realism, can be regarded as intelligible unless it obeys a definite criterion of ontological commitment".[1]

The honor of having formulated such a criterion goes to Quine. Although his criterion has aroused certain objections — which we shall examine — it has, however, won over the support of avowed Platonists like Church and radical nominalists like Goodman, which shows that he fully satisfies the requirement of neutrality.

Quine has formulated his criterion several times, most completely in 'Logic and the Reification of Universals' (1953) wherein he states: "*In general, an entity of a given sort is assumed by a theory if and only if it must be counted among the values of the variables in order that the statements affirmed in the theory be true*".[2] In other words, every existential commitment relates to quantified variables *to the exclusion* of predicates. Quine interprets these latter as syncategoremes devoid of autonomous meaning and ontological import. As J. Vuillemin stresses in *La logique et le monde sensible* (1971),

According to Quine, ... the four "open" statements: '$(\exists x)(x$ is red)', '$(\exists x)(\exists y)(x$ is parellel to $y)$', '$(\exists x)(\exists y)(x$ is like $y)$', '$(\exists x)(\exists y)(x$ is to the left of $y)$', commit us ontologically only to the individuals x, y and not to the property (or to the class) of being "red" or to the relations of "being parallel to", "being like", or "being to the left of".[3]

Thus, the following two sentences differ in what they assume:

$(\exists x)(x$ is a cell);
$(\exists \alpha)(x)(x$ is a member of $\alpha \supset x$ is a cell).

The first sentence assumes and affirms — we shall distinguish the two notions later — the existence of *individual cells*. The second assumes and affirms the existence of a *class* of cells.

CHAPTER I

In order to understand the notion of *ontological assumption* it is indispensable to distinguish it from related notions with which it has often been confused. To avoid any misunderstanding we must carefully explore the semantic field to which this key notion belongs. The multiple distinctions which we shall have to formulate precisely contain nothing pedantic or scholastic; they allow us to raise important new problems and to solve them to a certain degree.

In the first place, it is important to distinguish the *ontological assumptions* of a theory from the *ontology* of that theory. It would be wrong, Quine says, to identify

> the ontology of a theory with the class of all things to which the theory is ontically committed. This is not my intention. The ontology is the range of the variables .... But the theory is ontically *committed* to an object only if that object is common to all those ranges.[4]

Therefore, it is not correct to assert that the use of either of two contradictory sentences, such as '$(\exists x)(x$ is red$)$' and '$\sim(\exists x)(x$ is red$)$', carries the same ontological commitment. In fact, in order that the first sentence be *true*, red objects must exist, which is not the case for the second. These two sentences, therefore, do not make the same ontological assumption. However, they have the same reference, as their quantified variables have the same value-range, and two theories which differed only in respect of these two sentences would have the same *ontology*. The entities which are *ontologically assumed* by a theory are, therefore, a subset of the class of entities constituting the *ontology* of that theory.

Likewise, it is important to distinguish the *ontological assumptions* of an empirical theory from that which the classical definition of validity in logic obliges us to assume. It is not correct to affirm that a use of the sentence '$\sim(\exists x)(x$ is red$)$' obliges us to assume the existence of non-red beings by reason of the equivalence of this sentence with '$(x)(x$ is not red$)$', nor that '$\sim(\exists x)(x =$ Pegasus$)$' obliges us to assume the existence of beings distinct from Pegasus. Indeed, these sentences would remain true if nothing existed. They do not, therefore, carry any assumption, contrary to what K.H. Potter affirms in 'Negation, Names and Nothing' (1964).[5]

Consider, for example, the formula '$(x)(Fx \supset Gx)$'. This formula in itself does not carry any ontological assumption, since it is equivalent to '$\sim(\exists x)(Fx \cdot \sim Gx)$', which again could be true if nothing existed. On the other hand, if the author of the affirmation '$(x)(Fx \supset Gx)$' admits the classical theory of quantification, then, and for this reason alone, he must assume the existence of at least one individual. Indeed, the inference rules of classical quantification theory allow us to infer '$(\exists x)(Fx \vee \sim Fx)$'; and this carries an ontological assumption.

It is nevertheless necessary to separate carefully the ontological import of the schema '$(x)(Fx \supset Gx)$', which is nil when taken in *isolation* from the ontological import it *receives* from the classical theory of quantification; for there are non-classical quantification theories which formulate a definition of validity that also applies to an empty universe. These theories do not assume the existence of individuals.

Finally, it is important to distinguish the *affirmation* of existence from the *assumption* of existence. The affirmed content is always assumed, but the converse is not true. One may assume without affirming. For example, a use of the sentence '$(\exists x)(x$ is a dog)' affirms the existence of canine individuals, but not the existence of mammals. However, it assumes both if we take '(x) (x is a dog $\supset x$ is a mammal)' as analytical and if we use 'assumes' for 'implicitly asserts'. Likewise, when one affirms 'the class of beings that fly is not contained in the class of birds', that is:

$$\{x: x \text{ flies}\} \not\subseteq \{x: x \text{ is a bird}\},$$

one assumes, but does not affirm, the existence of an individual that flies but is not a bird. In fact, the truth of the above-quoted sentence implies the truth of

$$(\exists x)(x \text{ flies \& } x \text{ is not a bird}).$$

Finally, in order to dissipate one last ambiguity, note that that which is assumed by a sentence must be supposed to exist for the sentence *as a whole* to be true. The force of this remark appears clearly from the following example:

If there are infraviruses, they are invisible,

which means:

$$(x)(x \text{ is an infravirus} \supset x \text{ is invisible}).$$

The author of this statement, formulated within the framework of classical quantification theory, must assume neither the existence of infraviruses nor that of invisible beings. He assumes only the existence of beings who, if they are infraviruses, are invisible. Now, this is an *assumption* that does not commit him to much. In order to guarantee the truth of the statement in question, it suffices that he *assumes* the existence of hats and coats, for example, that is, a non-empty universe. As noted by Cheng and Resnick in

'Ontic Commitment and the Empty Universe' (1965), "it is possible for a theory to have ontic commitments without being committed to anything of which its primitive predicates are true".[6]

2. WARNOCK'S OBJECTIONS TO QUINE'S CRITERION OF ONTOLOGICAL COMMITMENT

Though Quine has virtually no competitors who offer a rival criterion of ontological commitment, he has, however, adversaries who contest the cogency of his criterion. The bearing that this has on our research requires that we thoroughly examine the criticism directed against this criterion.

There does not seem to be an alternative to Quine's criterion — though in these matters one must remain prudent and not prophesy — and it is not without interest to see why it is the only one of its kind. Signs are generally classified into variables and constants. *Individual constants*, that is, proper names, may be replaced in principle by definite descriptions, and these, in turn, may be eliminated. On the other hand, one may also eliminate *predicate constants* by treating them as syncategorematic signs not possessing an autonomous meaning. Consequently, one does not see exactly where *ontological commitment* can be localized, if it is not in the use of variables.

The only author who, to our knowledge, proposed a criterion of ontological commitment different from Quine's is Prior, and he does not exactly propose a rival criterion. He presents, rather, a criterion of ontological non-commitment. Prior thinks that an expression amenable to modal operators is free of ontological import; and one can find in this a contrast between Quine and Prior. For the former, quantified variables 'commit'; and for the latter, modal operators 'release'. The two criteria complement each other.

According to G. J. Warnock, Quine's criterion mixes metaphysical views with logic, views which are the more doubtful as their implicit character as a rule frees them from critical examination. Quine's criterion would be, in a way, a camouflaged metaphysical dogma. Warnock's essay devoted to Quine bears the eloquent title 'Metaphysics in Logic'.

Warnock summarizes Quine's doctrine on ontological commitment as follows:

We may discourse of classes, as Boole does, or of propositions, as in the propositional calculus, without thereby commiting ourselves to Platonism; for we can discourse in these ways without taking the fateful step of 'quantifying over' a class, or a proposition. If we take this crucial step, however, we fall into the ontological grip of the existential quantifier.[7]

THE CRITERION OF ONTOLOGICAL COMMITMENT 19

Quine, Warnock claims, also proposes to utilize the converse of existential generalization, that is, universal specification, as a criterion of ontological commitment. In other words, to know that a name 'a' is used seriously rather than fictionally, I must, for example, ask myself if this name alters the truth of the identity principle $((x)(x = x))$ when I substitute it for 'x'.

Warnock advances three arguments against these two versions of Quine's criterion:

(a) an existential reading of '$(\exists x)$' is not essential;

(b) verification of an existential statement does not require that one make claims about the nature of the objects in the value-range of the quantified variables;

(c) 'Pegasus = Pegasus' is true, whereas in terms of Quine's criterion it ought to be false, therefore Quine's criterion is inadequate.

(a) *The argument against an existential reading of '$(\exists x)$'*

In its everyday use, Warnock remarks, the word *something* has a function entirely different from that supposed by Quine: "One is ordinarily disposed to use the word 'something' in cases where one does not know what in particular, or where for some reason one does not wish to specify".[8]

It is clear that one may give a non-existential interpretation to the quantifier '$(\exists x)$', as suggested by Warnock. But one will then have to entrust to other signs the role of expressing our assumptions and affirmations of existence. All the same, if one deprives the natural language locution 'there exists some ...' of its existential sense, one will have to compensate for this by imposing, by fiat, on other locutions or signs the function normally performed by this locution, unless of course one is resigned to an arbitrary impoverishment of our means of expression. Therefore, Warnock's first argument is unacceptable.

(b) *The argument that the verification of an existential statement is independent of ontological assumptions*

Warnock argues using an arithmetical example; but what he says about numbers is supposed to apply to propositions also: "Dealing with integers we might say, 'For all values of x, $2x$ is even', or 'For at least one value of x, $x = 7 - 3$'."[9] If Quine is right, in uttering these statements are we not committed to the existence of abstract entities, that is, numbers? Are we not caught in the trap of the existential quantifier? In Warnock's mind this rhetorical question obviously calls for a negative answer:

... the existential statement in question, whether true or false, can be shown to be true or false by purely mathematical operations.... The matter is entirely remote from the arena both of Platonic and of anti-Platonic philosophizing.[10]

> The question whether there is an integer equal to 7 minus 3 is closed, once we have said and if necessary shown that 4 is such a number. Whether the number 4 itself exists is, if there could be any such question at all, a quite different question — a different *sort* of question; and certainly we do not answer it in the affirmative merely by answering affirmatively the other question.[11]

In support of this thesis, Warnock remarks that, "In fact every schoolboy knows quite well that it [For at least one value of x, $x = 7 - 3$] is true, even if he has never so much as heard of Plato ... .[12]

Warnock appeals here to the same distinction as Carnap in 'Empiricism, Semantics and Ontology' (1950),[13] where he distinguishes *questions internal to a science* from *external questions* concerning the conceptual framework of the science in question. Furthermore, the two works are contemporary. Like Carnap's, Warnock's distinction admits a pragmatism which minimizes the import of the controversy between realists and nominalists concerning the status of universals in general and propositions in particular. What he says about numbers applies, in fact, equally well to propositions. Besides, it is easy to imagine a new formulation of his argument in terms of propositions. Concerning the existential statement, 'There are propositions of astronomy which all philosophers of the Middle Ages accepted because they got them from Aristotle', Warnock would certainly say that we have here a metaphysically neutral affirmation concerning the history of ideas, decidable by methods employed by historians, which is far removed from the field where Platonists and anti-Platonists confront each other. He would no doubt contrast the sociological statement, '*Some* propositions are believed by all men' to the metaphysical statement, 'There *exist* propositions'.

Warnock, like Carnap, plays the card of pragmatism a bit short. Even if the internal (scientific) questions may be resolved *independently* of external (ontological) questions, it does not follow that the latter, as opposed to the former, may be *put in parentheses*, or resolved *conventionally* or *pragmatically*. The position of Quine, who considers that there is only a *difference in degree* and not a difference in nature between these two kinds of questions, avoids the *arbitrariness* which bedevils Carnap's distinction between internal and external questions, a distinction which is parasitic upon the analytic—synthetic dichotomy.

Although Warnock invokes the testimony of the candid schoolboy, we must recognize that such arguments are insufficient. Assuming and presupposing are *normative* notions. A reasoning may *require* a supplementary

THE CRITERION OF ONTOLOGICAL COMMITMENT

premise, even if a candid schoolboy does not feel the need of adding it. To deny this is to open the way to a *psychologism* that a logician cannot put up with.

Certainly, one may reason like a radical formalist and consider arithmetic a pure syntax which is closed in itself. But if one wishes to *apply* arithmetic to the real world, then the necessity of giving a semantic dimension to the variables will again arise along with the problem of ontological commitment that Warnock had finessed away. The schoolboy *must* assume that the numerals *designate* numbers, whether he *does* it or not. Since he does applied arithmetic, he assumes that numerals have a mathematical *meaning,* even if he is not *conscious* of it.

(c) *The argument against the criterion of exemplification*

Warnock then tries to 'trivialize' Quine's test by reasoning as follows:

Suppose we agree that, for all values of x, $x = x$; can we proceed to infer that appendicitis = appendicitis?
  This is not much help.... Why should such an expression as 'appendicitis = appendicitis' ever be written down, uttered, asserted, or denied? ... if we were for any reason persuaded to allow this sort of expression, it would be hard indeed for the speaker of plain English to see why any version of it should be, or indeed how it could be, denied. 'Pegasus = Pegasus' looks odd, but not deniable; there seems to be nothing wrong with 'pink = pink', nor yet with 'if = if'. If this is a test for designative use, then every expression designates.[14]

It is clear that in its liberality ordinary language, on which Warnock's view is founded, would also admit as 'undeniable' the identity $0/0 = 0/0$. The insufficiency of *linguistic impressionism* appears here very clearly. Indeed, the identity just mentioned would allow one to 'prove' that $2 = 1$, as Anderson and Johnstone have illustrated.[15] Here one sees the *unpalatable* consequences of Warnock's excessive confidence in ordinary language. The example shows that every valid specification of the variable x in an identity statement required the *univocity of the argument* substituted for the variable.

On the other hand, whoever accepts in the name of ordinary English that 'Pegasus = Pegasus', would also accept, 'The winged horse captured by Bellerophon = the winged horse captured by Bellerophon'. He would then be in conflict with Russell's theory of descriptions. Russell in *Principia* in effect *subordinates* the replacement of x in $x = x$ by a definite description to the existence of the entity so described.

$$14.28 \quad E! \, (\imath x)(\Phi x) \cdot \equiv \cdot (\imath x)(\Phi x) = (\imath x)(\Phi x)$$

Russell's and Whitehead's theorem formulates the idea that the substitution of a description for the variable *x assumes the existence* of the individual described. Admittedly, this theorem appears counter-intuitive and will count as an argument against Warnock only for those who are prepared to regiment language instead of describing it.

3. THE APPLICATION OF THE CRITERION OF ONTOLOGICAL COMMITMENT TO PROPOSITIONS

The main interest that Quine's criterion of ontological commitment has for us is that attributed to it by Church. This criterion allows us to formulate in *technically satisfying* terms the metaphysical problem of the status of propositions. Reformulated with the help of this criterion, the problem is stated as follows: *Do contexts exist where the word 'proposition' fulfills the function of a bound, that is quantified, variable and may not in this capacity be replaced by the word 'sentence'?*

Language can do without names, but it must contain variables which, by means of their values, link it to reality. The question that arises is to determine whether language must admit quantified *propositional variables*.

It is undeniable that Quine's criterion allows us to *circumscribe* the problem of the metaphysical status of propositions. In order to convince ourselves of this, consider the statements 1.1, 2.1, 3.1 and 4.1:

1.1   All propositions are identical with themselves.
1.2 $(p)(p = p)$
1.3 (s)(that s is the same proposition as s).
2.1   If one proposition implies another, then the negation of the second implies the negation of the first.
2.2   If p implies q, then $\sim q$ implies $\sim p$.
3.1 $(\exists p)(\exists q)(pq \cdot \exists \cdot p \sim q)$
3.2   '$pq \cdot \exists \cdot p \sim q$' is neither a logical truth nor a contradiction.
4.1   Albert conceals all the propositions that he believes.
4.2 (p)(A believe $p \supset$ A conceals p).

Before Quine formulated his criterion one would have said that the four statements expressed by these sentences presupposed the existence of propositions. But a different answer is possible. In the first sentence, in the natural language the word 'proposition' appears in the subject place, but there are two possible translations into formal language (1.2 and 1.3). According to the first, which is more onerous, the word 'proposition' is expressed by a quantified propositional variable. According to the second, the word 'proposition' is incorporated into the connective. Now, we know that connectives are considered expressions devoid of sense in isolation. Therefore, this second use of the word 'proposition' does not carry with it any ontological commitment to propositions, so long as we interpret the variable as a sentential variable rather than as a propositional variable.

Note that in 1.3 sentence variables are not combined with a dyadic predicate but are coordinated by a connective. In 1.2, variable 'p' admits names of propositions (nominalized sentences) as substituends (for example, 'that the earth is round') and *propositions* as values. In 1.3, the variable 's' admits sentences as substituends and takes as values whatever is chosen as the extension of a sentence — for instance, *truth values*; i.e. something other than a proposition.

In the second statement (2.1), the word 'proposition' in the natural language plays the role of a schematic letter. In fact, one may translate 2.1 as the formula 2.2 without affecting the sense of the former statement. Here again, despite appearances, one does not make an ontological commitment to propositions by affirming 2.1 or 2.2. In effect, a schematic letter is not a variable. One sees clearly here that Kneale was wrong in linking the generality of logical laws to the existence of propositions. As we shall see later on, one may transcend particular languages with schematic letters as well as with variables.

In the third statement (3.1), it appears at first blush that we need existential quantifiers. If we suppressed them, and if we treated 'p' and 'q' as schematic letters, we would be implying that '$pq \prec (p \cdot \sim q)$' is a theorem, which is not the case. We must therefore reject this claim of theoremhood; but, as Lewis and Langford have pointed out,

We cannot do this by asserting

$$\sim(pq \cdot \prec \cdot q \sim q)$$

because this affirms altogether too much.[16]

Indeed, the formula $pq \prec (p \sim q)$ is true for *certain* values 'p' and 'q'. For example, it is true if one substitutes '$\sim p$' for 'q' and keeps the other occurrences of 'p' intact. Lewis and Langford conclude that it is necessary to use the existential quantifiers in order to express this idea. Their conclusion, however, does not impose itself in any way. It is, in fact, possible to express the same thought metalinguistically by asserting that the formula '$pq \prec (p \sim q)$' is neither a logical truth nor a contradiction.

Consider now the fourth statement (4.1). This time it is no longer possible to give to the word 'proposition' in the natural language a role equivalent to that of a schematic letter in a symbolic language. In fact, in order to formalize statement 4.1, one must *quantify* over *p*. However, one cannot quantify over schematic letters. Only variables can be quantified. But we saw that it is through the use of variables that ontological commitments are made. The question of the ontological status of propositions is thus clearly circumscribed. To ask whether propositions exist is to ask if there exist contexts where propositions are the values of quantified variables, and *where this role cannot be denied them.*

4. COMPROMISING USES OF THE WORD 'PROPOSITION'

In order to consider this issue we must analyse the following four examples; the first two are from Ayer:[17]

(1a)   "My friend believes whatever he reads in the newspaper."
(2a)   "He asserted two propositions, which you probably believe but I doubt."

In a semi-formalized language the first sentence would be written:

(1b) $(\exists p)(p$ is affirmed in the newspaper$)$ & $(p)(p$ is affirmed in the newspaper $\supset$ my friend believes $p)$.

The second would be written:

(2b) $(\exists p, q)$ [(you believe p & I do not believe p & you believe q and I do not believe q & he asserted p & q & (r)(you believe r & I don't believe r & he asserted $r \supset (r = p) \vee (r = q)$].

To these two examples we may add the following ones:

(3a)   "I do not think of everything that I believe at every moment."
(3b) $(\exists p)(\exists t)$(I believe p at time t and I do not think of p at time t).

These formulable but unformulated beliefs seem, at first blush, to defy a translation into sentences.

(4)   'Albert who is not a polyglot, knows more truths than Bernard who is'.

In this example, the expeditious nominalist, who would replace 'truths' by 'true sentences' rather than by 'true propositions', would run the risk of falsifying the initial statement. Don't we have here a *case* where the role played by the concept of 'proposition' *cannot be* fulfilled by that of 'sentence', and don't we have a statement which, if formalized, would require the use of 'propositional variables' not replaceable by 'sentential variables'?

5. CRITIQUE OF AYER'S FIRST ATTEMPT TO ESCAPE ONTOLOGICAL COMMITMENTS TO PROPOSITIONS

In order to avoid the ontological implications of these cases of quantification over propositions, three manoeuvres have been attempted, the first two put forth by Ayer, the third by Warnock. We shall examine them one after the other.

In *The Foundations of Knowledge* (1940), Ayer affirms of such contexts:

> Nor need it give rise to any philosophical perplexity so long as we remember in using it that "meaning", "knowing", "believing" and the rest are not relations, like "loving" or "killing", that require a real object, ...[18]

As one can see, Ayer exploits the logical difference that separates *relational* from *intentional* verbs, as Husserl called them. It is necessary to distinguish carefully these two categories of verbs; indeed the *valid inferences* are not the same in these two cases. The rule of existential generalization (EG) may not apply to the direct object in the second case, whereas it does in the first. The *logical* difference is well known. 'I killed a person', implies that a victim exists; 'I am looking for a treasure' does not imply that a treasure exists. Those who *do not recognize* this difference may reason incorrectly.

Like Warnock, Ayer attends only to the ontological import of the *existential quantifier*. He reasons as though ontological commitments were exclusively revealed by the existential quantifier. Yet one of the merits of Quine's criterion of ontological commitment was to make us aware of the ontological import of *bound variables*, including those occurrences bound by universal quantifiers.

Ayer wrote on the subject in 1940 and 1947. In 1953, Quine formulated his criterion of ontological commitment. In 1939, in 'Designation and Existence', he had already written: "To be is to be the value of a variable".[19] But in this paper the contours of the criterion of ontological commitment, which are so clear today, were blurred *as a result of their proximity to another criterion*:

> To say that there is such a thing as appendicitis, or that "appendicitis" designates something, is to say that the operation of existentially generalizing with respect to "appendicitis" is *valid*; i.e., that it leads from truths only to truths.[20]

Now, this other criterion, that of the applicability of existential generalization, coincides precisely with the one Ayer uses to distinguish relational from intentional verbs. It is therefore reasonable that Ayer, when pondering this text, was sensitive to one of Quine's two criteria which intersected his own.

Quine's two criteria are fundamentally different. The one formulated in terms of existential generalization is a syntactic criterion. The rule of existential generalization is in effect an inference rule from logical syntax, a rule which conceals a presupposition of existence. Hintikka has shown that this rule is independent of the other inference rules of quantification theory.[21] One may thus abandon it, and then restore it by making its use subordinate to an appropriate premise of existence. One consequently obtains a logic of quantification without the presupposition of existence.

The second criterion, the one that today Quine cites exclusively, is *semantic*. It affirms that a variable has *meaning*, that is, reference, to use Frege's terminology, only if one assigns to it a non-empty value range. Quine, himself, according to Martin, would occasionally have violated this fundamental rule.[22]

It is useful to *distinguish* Quine's two criteria, because this distinction reveals to us the inadequacy of Ayer's solution, which simply consists in subdividing the class of verbs into a subclass of relational verbs and a sub-class of intentional verbs, and in pointing out that the latter do not presuppose the existence of entities designated by their direct objects.

This initial parry from Ayer only meets one of the two exegencies of Quine's criterion. It certainly allows its author to *avoid invalid reasonings* that would have *produced affirmations of existence* relative to propositions, but it does not avoid the affirmation of existence involved in the use of quantified propositional variables, where by chance these affirmations were not obtained by existential generalization. In other words, Ayer's first parry is effective only against the *syntactic* danger of intentional verbs; but it does not allow him in return to *honor the semantic obligation* he himself creates by using bound variables ranging over propositions.

6. THE DOUBLE INTERPRETATION OF THE EXISTENTIAL QUANTIFIER

In *Thinking and Meaning* (1947), Ayer takes up the preceding argument and combines it with a new idea: there could be *several possible readings* of the existential quantifier. One would be ontologically compromising ('$\exists x$' = 'there *exists* an x which'); the other would be ontologically neutral ('$\exists x$' = 'there is an x which'). He writes:

'To be sure, it makes sense to say, in a case where someone is believing or doubting, or whatever it may be, that there is something that he doubts or believes. But it does not follow from this... that something must exist to be doubted or to be believed, ...'[23]

We note in passing that Ayer does not subscribe to the disastrous thesis of the double sense of being, but, rather, to a quite different one; namely, the thesis according to which there are ontologically neutral and ontologically committed uses of 'there is'.

Contrary to what one might think at the outset, the way of avoiding ontological commitments put forth by Ayer is not open to the objection levelled against the non-existential interpretation of the quantifier '$(\exists x)$' proposed by Warnock. In fact, Ayer does not deny himself the possibility of expressing ontological commitments in this way. He retains, in effect, the classic conception of the existential quantifier, but adjoins to it a quantifier which does not carry with it ontological commitment. Is this *ad hoc* solution acceptable? Before answering this, one must first ask whether it is possible to allow distinct readings of the existential quantifier *in the same language* without producing an inconsistent logic.

This latter objection was formulated by Church as follows: "... it has to be said that neither a logic nor a semantics of this latter notion of existence has even been indicated vaguely, much less received precise formulation. The notion remains mysterious".[24] Church wants to see a properly formalized language in which one can handle two existential quantifiers without provoking a catastrophic collapse of the two notions. Unless one succeeds in constructing such a calculus, Ayer's *distinction* is not *operational* for it does not modify in depth our manner of reasoning.

Church's demand was recently met. In *A Translation Theorem for Two Systems of Free Logic* (1967), Lambert and Scharle have shown that it is possible to handle, *without incoherence*, two different existential quantifiers, one of which has ontological significance and the other not. Thus, one may write without incoherence:

$$(\exists x)(x \text{ is a proposition}) \cdot \neg (\mathsf{E} x)(x \text{ is a proposition});$$[25]

that is, 'There are propositions and no propositions exist'. Note that this calculus makes use of a *non-reflexive* notion of identity axiomatically defined by Lejewski in 1965.

In the light of the controversy between Ayer and Church, the technical results obtained by Lambert and Scharle receive a new significance. They confer on Ayer's argument the technical support it needs to satisfy the exigencies formulated by Church. This shows that by stating his objectives *precisely* a philosopher may, as he did in this case, completely resolve certain problems.

We recognized that the argument by means of which Ayer tries to relieve the existential quantifier of its ontological import can be substantiated, and that one of the grounds for assuming the existence of propositions is thereby eliminated. But the question of whether one *must* assume the existence of propositions is not yet solved, for there are contexts where we must apply the existential quantifier in its classic sense to propositional variables and where Ayer's manoeuvre is inoperative.

In these contexts, however, it is possible to escape ontological commitment by other means; instead of modifying the semantics of the *existential quantifier* (by changing its sense) or its syntax (by changing the rules of inference concerning it) one may change the type of the *variable*. One may construe the variables 'p', 'q' as *sentence* variables, that is, so that they take sentences as *values*. One may also, more fundamentally, isolate the variables 'p', 'q' from their values and retain only their substituends, which are in any case sentences and not propositions. We shall examine these two manoeuvres in turn.

7. THE DOUBLE INTERPRETATION OF BOUND VARIABLES

In 'Propositions and Abstract Propositions' (1968), Colwyn Williamson recognizes that one can avoid the use of quantified variables to report the intentions and beliefs of a third. For example, formalizing the statement,

> Everything the policeman said was true,

one would have:

> (p)(if the policeman said p, then p is true).

But, as Williamson remarks, these variables need not necessarily be interpreted as *propositional* variables:

Cases like "What the policeman said was true" prove nothing about the nature of propositions; they merely show the need for (or, more exactly, illustrate the role of) variables, or their ordinary language equivalents. We may, if we wish, call them "sentential variables"; or we may call them "propositional variables" and go on to give an account of propositions which identifies them, in one sense or another, with sentences.[26]

Thus, Williamson adopts the first solution to which we have already alluded. This interpretation of the variable 'p' as a variable taking sentences as values, unfortunately does not harmonize with the theory of schematic letters. A logic which simultaneously adopted Williamson's interpretation of variables and Quine's of schematic letters would be equivocal; '*p*' would be treated as a name of a sentence and as indicating a *place* to be occupied by a sentence. The second of these two interpretations, that is, the substitutional interpretation, on the contrary seems to us to harmonize perfectly with the theory of schematic letters and even to extend it.

In 'Oratio Obliqua' (1963), Prior tends toward this second interpretation:

... like many other difficulties of his [Quine's], it [the difficulty raised by quantification within belief-sentences] only arises through his insistence that all quantification must be over variable *names*. I see no reason why we should not — still following Ramsey — quantify over *sentential* variables and concoct such complexes as "For some *p*, Paul believes that *p* and Elmer does not believe that *p*," without thereby being "ontologically committed" to the view that there are objects which sentences *name*...[27]

It is quite true, as Prior states, that the ontological import that Quine attributes to variables stems from the fact that he treats variables like *indeterminate names*; individual variables being indeterminate names of individuals, class variables being indeterminate names of classes, and, consequently, propositional variables, indeterminate names of propositions. It is likewise true, as we have already intimated, that this interpretation is not the only possible one. *One may treat variables not as indeterminate names having entities as 'values', but simply as indeterminate signs for which one may substitute determinate signs, without raising any question about what these signs stand for.*

This second solution, initiated by Lesniewski, has been elaborated by R. Barcan Marcus in another context.[28] Quine calls this interpretation of the variable a *substitutional* interpretation, as opposed to the classical *objectual* interpretation. In *Ontological Relativity and Other Essays* (1969), he defines the substitutional interpretation of variables bound by an existential quantifier as follows:

An existential substitutional quantification is counted as true if and only if there is an expression which, when substituted for the variable, makes the open sentence after the quantifier come out true. A universal quantification is counted as true if no substitution makes the open sentence come out false.[29]

In other words, for Barcan, '$(\exists x)(\Phi x)$' does not mean 'There are some *values* of '*x*' for which 'Φx' is true', but 'There are some *substitutions* for '*x*' for which 'Φx' is true'. This second interpretation is basically different from the first, for the *value* of a variable is generally something extra-linguistic, where what is substituted for a variable is a *sign*, thus a linguistic entity.

For the nominalist, therefore, the Lesniewski-Barcan-Prior interpretation of the variable is attractive. Why did Quine, who is inclined toward nominalism, not adopt it? The reason is very simple. When we remove any ontological import from variables bound by existential quantifiers, we must confer on other signs the function of anchoring our language in reality; for if not, we would risk remaining incapsulated within language, as we have already pointed out in examining the interpretation of the existential quantifier proposed by Warnock. Onto which signs are we to confer this anchoring role? They cannot be predicates, since it is agreed to treat them as syncategoremes. Obviously, we can only turn to proper names, which is what Wittgenstein does in the *Tractatus*. But it is precisely this solution that Quine does not want; on the contrary, he means to eliminate proper names in favor of bound variables.

There are good reasons for not adopting *in all cases* an interpretation of quantified variables that obliges us to push back onto names the functions fulfilled by variables in the *classical* interpretation. In certain cases, we do not, in fact, *have* enough proper names to allow us to replace the existential quantifier by a list of such names. Not all celestial bodies, for example, have a name. For this reason we could not replace the quantifier of '$(\exists x)(x$ is a celestial body$)$' by a list of names. In other cases, not only do we not have enough names, but we *could not* have enough, even if we allowed ourselves an infinite list of them. This is the case if we wished to express without the help of classical quantification what is expressed by '$(\exists x)(x$ is a real number$)$'. As is well known, linguistic signs which are possible substituends for variables form a denumerably infinite set, whereas the value range for class variables is a non-denumerably infinite set — as Cantor has shown in his set theory. The unilateral adoption of Barcan's interpretation would, therefore, oblige us to reject a fundamental result of this theory.[30] This is a sacrifice which a nominalist may not agree to make, if, like Quine and ourselves, he subscribes to Stegmüller's methodological principle quoted above (Introduction § 4).

It was shown why one could not, in general, subscribe to the substitutional interpretation of variables advocated by Barcan; but it is important to remark that the arguments set forth by Quine do not exclude a *piecemeal* adoption of Barcan's interpretation. Why could one not keep the classical interpretation of variables for individual variables and class variables, and adopt the Barcan-Prior interpretation for propositional variables?

The *differential* treatment of variables we suggest may seem to be *ad hoc*. It is undeniably a manoeuvre which is intended to protect doctrinal nominalism at the expense of the unity of theory. If we subsequently find a means to provide a *homogeneous* treatment of quantification, we shall have to prefer this solution to the present compromise.

Thus, we propose to adopt *provisionally*, and as a working hypothesis, the *mixed* interpretation of variables. Before definitely accepting this solution, we shall have to see if it accords with the facts revealed in the study of belief and indirect discourse.

Note, however, that if one subscribes to the definition of the proposition proposed by Jan Berg in 'What is a Proposition?' (1967),[31] then one cannot adopt the substitutional interpretation of quantification. Indeed, Berg's definition is presented in the context of a theory that posits as an axiom that there is a *non-denumerable* infinity of propositions. Such an axiom clearly derives support from the following consideration: if there is a non-denumerable infinity of real numbers, there is a non-denumerable infinity of propositions about real numbers. Berg's argument is compelling. One cannot *a priori* legislate that the set of *propositions* is denumerable. The set of sentences, however, certainly is since the set of well formed expressions has to be recursive.

8. FROM PRAGMATICS TO ONTOLOGY

The criterion of ontological commitment has implications for the *pragmatics of language*. It allows us to uncover the philosopher's conscious and unrevealed ontological commitments, and as such has great dialectic utility. It may provide an efficient tool in the discussion by revealing that some followers of nominalism must nevertheless subscribe in practice to positions which they reject in theory. But contrary to what is generally believed, its utility does not stop here. The pragmatic criterion of ontological commitment may give rise to an ontological criterion, properly speaking. At the outset, however, there is none. When Quine affirms that "to be is to be the value of a variable", he does by no means answer the question which, for example, Berkeley answers when he affirms that *"esse est percipi"*. In fact, Quine's *aphorism* is an *elliptic* affirmation. Completely formulated, Quine's principle would be stated as follows: 'To *treat* objects of a certain category *as* values of a variable is to *assume* the existence of these objects'.

Having made these reservations, one must admit that it is easy to transform the criterion of ontological *commitment* into an ontological criterion, properly speaking. In order to confer onto Quine's criterion this ontological prerogative, it suffices, we believe, to *combine* it, for example, with Russell's law of parsimony to which we subscribed at the beginning of this work. In fact, if we say, conforming to the spirit of this law, that the entities which we cannot avoid assuming *effectively exist*, and if, thanks to the criterion of ontological commitment, we see that we *must assume* certain entities as values of our variables, in order not to impoverish the discipline we must conclude that these entities actually exist.

REFERENCES

[1] A. Church, 'Ontological Commitment', *Journal of Philosophy* 55, (1958) 1012.
[2] W. V. O. Quine, *From a Logical Point of View*, Harper Torchbooks, 2nd edn., 1961, p. 103.
[3] J. Vuillemin, *La logique et le monde sensible*, Flammarion, Paris, 1971, p. 43.
[4] W. V. O. Quine, 'Replies', *Synthese* 19 (1968) 287; repr. in D. Davidson and J. Hintikka (eds.), *Words and Objections*, Reidel, Dordrecht, 1969, p. 315.
[5] K. H. Potter, 'Negation, Names and Nothing', *Philosophical Studies*, (1964) 52.
[6] Ch. Cheng and M. Resnik, 'Ontic Commitment and the Empty Universe', *Journal of Philosophy* LXII, (1965) 361.
[7] G. J. Warnock, 'Metaphysics in Logic', *Proceedings of the Aristotelian Society, 1950–1951*, repr. in A. Flew (ed.), *Essays in Conceptual Analysis* MacMillan, London, 1956, p. 87.
[8] Warnock *Ibid.*, p. 81.
[9] Warnock *Ibid.*, p. 87.
[10] Warnock *Ibid.*, p. 87.
[11] Warnock *Ibid.*, p. 88.
[12] Warnock *Ibid.*, p. 87.
[13] R. Carnap, 'Empiricism, Semantics and Ontology', *Revue Internationale de Philosophie* (1950); repr. in *Meaning and Necessity*, 2nd edn., 1956, pp. 205–221.
[14] G. J. Warnock, *op. cit.*, p. 84.
[15] J. M. Anderson and H. W. Johnstone, *Natural Deduction: The Logical Basis of Axiom Systems*, Wadsworth Publ. Co., Belmont, 1962, p. 241.
[16] C. I. Lewis and C. H. Langford, *Symbolic Logic*, Dover, New York, 1932; 2nd edn., 1959, p. 184.
[17] A. J. Ayer, *The Foundations of Knowledge*, MacMillan, London, 1940, p. 103.
[18] Ayer, *Ibid.*, pp. 103–104.
[19] W. V. O. Quine, 'Designation and Existence', *Journal of Philosophy* 36, (1939); repr. in H. Feigl and W. Sellars (eds.), *Readings in Philosophical Analysis*, Appleton-Century-Crofts, New York, p. 50.
[20] Quine, *Ibid.*, p. 48.
[21] J. Hintikka, 'Existential Presuppositions and Existential Commitments', *Journal of Philosophy* 56, (1959) 132.
[22] I. Martin, 'Existential Quantification and the Regimentation of Ordinary Language', *Mind* 71, (1962) 528.
[23] A. J. Ayer, *Thinking and Meaning*, H. K. Lewis, London, 1947, p. 14.
[24] A. Church, 'Ontological Commitment', *Journal of Philosophy* 55, (1958) 1010–1011.

[25] K. Lambert and T. Scharle, 'A Translation Theorem for Two Systems of Free Logic', *Logique et Analyse* 39–40, (1967) 339.
[26] C. Williamson, 'Propositions and Abstract Propositions' (1968) in N. Rescher (ed.), *Studies in Logical Theory*, Blackwell, Oxford, 1968, p. 143.
[27] A. N. Prior, 'Oratio Obliqua', *Proceedings of the Aristotelian Society*, Supplementary Volume 37, (1963) 118.
[28] R. Barcan Marcus, 'Modalities and Intensional Languages', in M. W. Wartofsky (ed.), *Boston Studies in the Philosophy of Science* Reidel, Dordrecht, 1963, pp. 91–93.
[29] W. V. O. Quine, *Ontological Relativity and Other Essays*, Columbia University Press, New York, 1969, p. 104.
[30] W. V. O. Quine, 'Reply to Professor Marcus', Wartofsky, *op. cit.*; repr. in Quine, *The Ways of Paradox*, Random House, New York, 1966, pp. 175–182.
[31] J. Berg, 'What is a Proposition?' *Logique et Analyse* 39–40, (1969) 283–292.

CHAPTER II

THE SYNTACTIC APPROACH

1. IS AN AXIOMATIC DEFINITION OF PROPOSITION POSSIBLE?

Let us briefly recall what one understands by the term 'axiomatic definition'. The axioms of a discipline contain *primitive terms*. They allow any interpretation compatible with the truth of the axioms in which they occur. By the same token they exclude other interpretations. One may thus, with Tarski, assimilate axioms to propositional functions and primitive terms to variables. Axioms effect a sorting-out among the objects that belong to the domain constituting the value-range of these variables; they select the classes of those objects of which they are true. It is this power of selection, of delimitation, that one associates with *definition*.

Although they do not provide expressions which are everywhere and always substitutable for primitive terms, axioms are nevertheless definitions to the extent that they delimit and circumscribe the classes that constitute the extension of primitive terms. Thus, for example, the arithmetician countenances a universe containing an infinitude of *any* objects, and, by means of postulates of arithmetic in which the word 'number' occurs as a primitive term, he 'filters' out from this set a subset of objects which are no longer *any* objects: the numbers. It is, therefore, legitimate to answer the question 'What is a number?', by saying, 'It is everything that satisfies the postulates of arithmetic', non standard models being set aside.

Once this bridge is crossed, one will naturally ask oneself if *proposition* could not be defined in the same way as *number* in the following terms: 'A proposition is anything that satisfies the primitive propositions of a certain calculus'. As attractive as it may seem, this extension to logic of a method of definition that was successful in arithmetic does not work. For, though resembling arithmetic in having the character of a deductive science, *logic* differs from it, as it also differs from all deductive sciences, in that "logic... does not presuppose any preceding discipline...".[1]

On examination, this 'priority' of logic will appear as an insurmountable obstacle for the axiomatic definition of the notion of proposition. This difficulty was pointed out in 'Appendix C' to the second edition of *Principia Mathematica* (1927), and it was remarked that it was likely to embarrass the philosopher rather than the logician:

Thus from the formal point of view ... it does not matter what propositions are, so long as we are content to regard our primitive propositions as defining hypotheses, ... (From a philosophical point of view, this formal procedure may be shown to presuppose the non-formal interpretation of our primitive propositions; but that does not matter for our present purpose.)[2]

The anomaly presented by an axiomatic definition of the proposition rests in the impossibility of keeping sufficient distance from the axiomatized notion in order to let the axioms play their defining role. In effect, the notion of a proposition that one seeks to define with the help of axioms is in a certain way presupposed by the statement of the latter, in the sense that axioms are themselves propositions.

Is there a *vicious circle* or merely an acceptable *circularity*, such as appears in the fact that one has to use energy in order to define energy? In our case, it seems quite clearly to be a vicious circle; for, while energy is what it is without human intervention, the elements of a calculus have the status of propositions only due to an implicit or explicit convention. That is certainly the opinion expressed by Quine in the following passage:

It is customary to consider systems in abstraction from the nature of their elements; the theorems of a system, thus viewed, become sentences telling us various properties of unidentified elements. But to abstract from the fact that the elements of the propositional calculus are propositions is to deprive the theorems *themselves* of their character as sentences, since in that calculus the theorems are symbols of elements of the system. The student of systems in the abstract thus comes to an *impasse* when he takes up the calculus of propositions.[3]

One may illustrate Quine's observation by the following example. Consider the theorem of the propositional calculus 'p or not-p'. If we refuse to consider 'p' a variable whose substituends are propositions in order to construe it as a variable whose substituends are indeterminate elements, the theorem 'p or not-p' itself is no longer a proposition.

2. TWO NOMINALIST SOLUTIONS TO THE PROBLEM OF INTERPRETING PROPOSITIONAL VARIABLES

In his 1934 article, Quine came very close to the theory of schematic letters, but the concern to provide *denotations* for the variables 'p', 'q', etc., points in another direction. Instead of looking for a new way of defining proposition, he attempts to free himself of this recalcitrant notion. Our present purpose requires that we pause for a moment and examine this undertaking and its extensions. Its most striking feature is that variables which were previously *propositional* become *sentential* variables having the class of *sentences* as their value-range.

The signs '*p*', '*q*', etc., thus become sentence variables: neither signs ambiguously denotative of propositions nor signs ambiguously abbreviative of sentences, but signs ambiguously denotative of sentences.[4]

The connectives no longer link propositions, nor even sentences, but are now attached to *names* of sentences, and are signs for the operation of inserting into sentences such and such adverbs and conjunctions. For example, the expression '*p* ⊃ (*p* ∨ *q*)' does not *denote* propositions, does not *abridge* sentences in the manner of a *shorthand notation*, but denotes sentences. It is a *name*. To transform this into a sentence, it must be preceded by the assertion sign (⊢), which here plays the part of a predicate attributing truth to the denoted element which appears, as Quine says, "in its wake".

Thus, in 1934, Quine along with many others shared the opinion that the semantic relationship linking propositional signs to that which they signify could be conceived on the model of the relationship that unites a name to the thing named. Variables do not distinguish, they are simply ambiguous names denoting entities unified in their value-range. Thus, Quine's nominalism essentially consisted in the fact that he granted *sentences* and not *propositions* the status of objects which are denoted or designated by propositional variables.

Today, Quine no longer conceives the meaning of propositional variables on the model of *denotation*. Presently, he conceives of it on the model of *schematization*:

We can view '*p*', '*q*' etc. as schematic letters ...; and we can view '[(*p* ⊃ *q*) • ~ *q*] ⊃ ~ *p*' ..., not as a sentence but as a schema or diagram such that all actual statements of the depicted form are true.[5]

This is not, as one might believe, to adopt a conception of variables as *abbreviations* of sentences, a view to which Quine had already alluded in his 1934 paper. The schematic letters are by no means *abbreviations*, but *blanks, voids, dotted lines*.

The replacement of propositional variables by schematic letters is a technical innovation weighty in a philosophical sense which we shall try to unravel. The difference between abbreviations and schematic letters may, we believe, be summed-up in the following contrast: schematic letters or dotted lines have the same *operative virtues* as the former *variables*, whereas an abbreviation is only a simple typographical artifice.

Not counting the fact that it allows us to dispense with the rule of substitution, schematic letters still differ from variables in the following two ways:

(a) The relation between a schematic letter and the things schematized differs from the relation between a variable and its values;

(b) Schematic letters are not amenable to quantification.

The *resemblance* between schematic letters and variables poses no problem. We shall therefore concentrate on the *differences*.

(a) A variable is an 'indeterminate name'. It is linked to the constants in its value-range by the *semantic* relation of *denotation*. A schematic letter is linked to what it schematizes by a *syntactic* relation: it abstracts the structure common to *several* sentences. In other words, while the variable is related to the objects in its range by the *semantic* relation of name to object named, the schematic letter is related to what it schematizes by the *syntactic* relation linking a *variable* to its *substituends*. Thus Quine states,

A schema such as '$(x)(Fx \supset p)$,' ... is not a name of a sentence, not a name of anything; it is *itself* a pseudo-sentence, designed expressly to manifest a form which various sentences manifest. Schemata are to sentences not as names of their objects, but as slugs to nickels.[6]

(b) A consequence of this difference in status is that quantification with respect to schematic letters makes no sense, while it is of course, meaningful to quantify over variables. This inaccessibility of schematic letters to quantification is very important philosophically. In effect, as we have seen in Chapter I, the ontological assumptions of discourse evaluated in terms of Quine's criterion of ontological commitment are exclusively concentrated on the quantified variables. Since schematic letters are not quantifiable, they are deprived of all ontological import. The problem of universals thus receives a *more precise formulation*: the question, 'Must one assume propositions?' may be restated as, 'Can one not replace propositional variables by schematic letters?'

3. WHAT QUINE'S NOTATION REVEALS WITH REGARD TO THE STATUS OF PROPOSITIONS

The transition from Quine's first conception to his second seems to indicate a *notable reinforcement of his nominalism*. Whereas propositional variables are susceptible of several interpretations — they can denote propositions as well as sentences — dotted lines, on the contrary, do not give us any choice;

they can only be replaced by linguistic expressions. Any necessity of a return to *Platonism* is therefore averted from the start.

Quine's schematism has another, less immediately perceptible consequence which is worthy of examination. So long as one employed propositional variables, one was forced to assume the existence of a fundamental *propositional* level, that of unanalyzed propositions. This suggested the idea that there exist *atomic* propositions which are themselves not again decomposable into propositions, though they might be so analyzable into smaller units other than propositions. Without doubt, logical atomism does not have to receive an *ontological* interpretation, as one may — and Granger has done — liberate logical atomism from its realist ties, formulating it in the following terms:

It is necessary that something *functions as* a simple object so that propositions may be laid down as elementary; it is necessary that some propositions *function as* elementary propositions so that the construction of compound propositions may be undertaken.[7]

It is nevertheless a fact that, for its defenders, logical atomism is often associated with a certain realism, as is shown in the case of Wittgenstein, who spent some time separating the two doctrines.[8]

Now, it is interesting to note that the danger of seeing atomism entail realism is completely eliminated as soon as one uses schematic letters in place of variables. And this danger is eliminated not only for the obvious reason that if propositions disappear, atomic propositions do likewise, but also for another reason which is non trivial. In Quine's 'schematism', interest in the *entities* joined by logical constants gives way to an interest in the *relations* of *resemblance* and *difference* among these latter; for knowledge of these relations is all that is needed in practice in order to fill empty places in the schema. The applicability of logical syntax no longer presupposes the existence of atomic propositions.

It is remarkable that, by means of a shift in perspective brought about in this way, the constraints of logic become more natural. This becomes convincing when the following is considered: To know that the present reasoning is valid, I do not need to settle the question of whether the first line contains *one* or *two* propositions; I do not even have to suppose that this question makes sense and that there *exists* a *determinate number* of propositions in the first part of this argument:

> 'The earth is round and the moon is a satellite; therefore, the earth is round and the moon is a satellite or Venus is a planet.'

On the other hand, I must know that this reasoning exemplifies, bearing in mind the resemblances and differences, the following 'macroscopic' structure:

$$\begin{array}{c}\therefore \quad \ldots \quad \vee \text{---}\end{array}$$

which is that of the rule of addition and which *suffices* to guarantee its validity.

Russell contrasted the logical with the grammatical form of propositions. This was an important step in the emancipation of logic from grammar. But in speaking of *the* logical form, as though there were only one for each sentence, Russell still preserved a certain realism, from which we were freed by Quine's 'schematism'. It is certain that an absolute notion of logical form is not an operational one. We can analyze reasoning in different ways, and we can give finer and finer analyses. But the limit which we seek is not that of *intrinsic atomicity*; rather, it is that of the minimum analysis required to guarantee the validity of the reasoning that one has formalized. The limit is not so much in the logical object as in the *operation* of validation.

When formalizing an argument in natural language, we must extract as many logical structures as are needed to validate the argument, in the way a driller sinks his shaft until he reaches ground-water. It is the same for auxiliary premises. One adds to the reasoning as many of these as necessary; and one appropriates them, of course, from among previously demonstrated theorems. Obviously, in proving the *existence* of an *optimum* in the analysis, we have not yet proven the *non-existence* of a maximum; that is to say, the inexistence of propositional atoms refractory to all further analysis. We have, however, at least deprived the realist of the support he had hoped to find in logical syntax and thereby displaced the burden of proof.

In a formalized system, the distinction between complete and incomplete proofs is clear. The same is true for the distinction between complete and incomplete axiom systems. In return, the designation of enthymeme attributed to an argument expressed in a natural language presupposes an analysis, a formalizing of this argument, which itself is not formal. In certain cases, an argument will or will not be an enthymeme, a statement will or will not be elliptical, depending upon how one formalizes them.

The philosophy which inspired Quine's notational innovations becomes clear when one compares it to that of the later Wittgenstein's followers; for they too wanted to eliminate the realist residue still present in Russell's conception of logical form. In connections with the followers of the later Wittgenstein, R. Wells writes in 'Meaning and Use' (1954):

Russell paraphrases 'Tame tigers growl' first as 'All tame tigers growl' and then, more analytically, as 'All things that are tame tigers growl'. 'Tame tigers exist' he paraphrases as 'There are tame tigers'. So far the Wittgensteinians would agree with him. But in going further and holding that the final paraphrases more accurately represent the real logical form, he makes a metaphysical commitment which they regard as unnecessary.[10]

4. DOES THE DEFINITION OF LOGICAL TRUTH PRESUPPOSE THE CONCEPT OF PROPOSITION? STRAWSON'S THESIS

The theory of schematism outlined above tends to reduce identity between propositions to typographical identity. The clarity, the radical nature of this conception, makes it more amenable to a rigorous critical examination. We shall now try to determine the significance of Strawson's critique.

Quine calls logically true those schemas which remain true for all replacements of their blanks by statements. Strawson maintains in 'Propositions, Concepts and Logical Truths' (1957) that "...Quine's characterisation of logical truth can be made coherent, and made to do its job, only by implicit use of notions belonging to the group which he wishes to discredit";[11] that is, by using such notions as those of *proposition, concept* and *synonymy*. In other words, Strawson claims that we cannot define logical truth without appealing, if only laterally, to the concept of proposition. He defended this idea again in 'Paradoxes, Posits, Propositions' (1967).[12]

For Quine, a truth of logic is a sentence which is a substitution instance of a schema, all substitution instances of which are true. But these interpretations and substitutions must be *uniform*. Now, Strawson remarks that typographical uniformity, though necessary — (doesn't one say of logical truths that they are true by virtue of their form?) — is, however, not sufficient given the existence of *homonymy in natural language*. Thus, for example, whoever blindly applied Quine's test ought to conclude that the statement: 'If he was seen by the bank (river), then he was seen by the bank (commercial)' is a logical truth because it has the form '$p \supset p$'. And this is absurd.

To avoid this disastrous consequence, one must admit, Strawson believes, that the application of a uniformity rule depends on the prior existence of a propositional identity. To subordinate propositional identity to typographical identity would in his eyes mean taking the effect for the cause: "Surely the point of any such rule, whatever its form, must be to preserve some kind of identity *already present* in the components for which the substitutions are to be made".[13] To remove the above-mentioned counter-example and other less important ones, Strawson believes that the definition of logical truth cannot do without an allusion to 'concepts' and 'propositions', as opposed to 'words' and 'sentences' and to *typographical resemblances*.

To show more clearly the inevitability of this reference to propositions, Strawson tried to foresee the nominalistic replies and to criticize in advance the adjustment nominalists would make to save their doctrine. Thus he tried to show that the effort to define logical truth in nominalistic terms had to fail completely, even were one to allow *nominalism* to be flexible enough to countenance truth-values.

Suppose one said that typographical uniformity does not serve to reflect synonymy, propositional identity, but simply extensional identity — that is, identity of the truth-values of sentences. This would incontestably be a less onerous explanation than that which draws on the stock of suspect notions of proposition, concept and synonymy. But Strawson maintains that this economical solution is not admissible:

If it were no more than this, we should do better to drop the symbolism of '$p \supset p$' in favour of '$T \supset T$' and '$F \supset F$', to which there would then be no reason for not adding '$F \supset T$'. We should then, as far as the propositional logic is concerned, be in the position of counting every truth-functionally compounded *truth* as a logical truth — which is certainly not the intention of those who speak of logical truth.[14]

This challenge addressed to Quine to define logical truth without mentioning the word 'proposition' remained unanswered at the time, though it did not go unnoticed;[15] and in 1967 Strawson took up the same argument and developed it:

If identity of truth value were all that recurrence of sentence-letter represented, it would be more candid to rewrite logic, with only two sentence-letters, "F" and "T", so that "$F \supset T$" was included among the forms of logical truths. The point is that we must be able to be sure that identical sentences have the same truth value independently of knowing what that truth value is. So back to constancy of interpretation.[16]

5. REPLIES TO STRAWSON'S OBJECTIONS

Strawson's claim, according to which *sense* is prior to *sign*, is an indisputable truth. The case of the bad poet for whom 'sense follows rhyme' no doubt illustrates a pathology of language. What seems less sure to us are the antinominalistic conclusions, or certain of them, that Strawson draws from this evidence. There are three conclusions which I shall order according to their decreasing credibility.

The first is formulated as follows: the correct use of an argument schema to represent an argument formulated in natural language requires that one replace schematic letters or similar variables in the schema by typographically similar and univocally-interpreted sentences. This assertion is banal and need not delay us further.

Strawson's second claim is grafted onto the first. It is much more interesting, but also much less evident. It consists in recognizing a priority of the sense of a sentence or predicate over the sign, that is to say, over the sentence variable or predicate variable, which would confirm the prior existence of *propositional* or *conceptual invariants*. If it is incontestable that the typographical identity of the sign reflects the identity of the signified content, it does not follow, I believe, that there exists *separate* propositional invariants or even simply *distinct* propositional invariants.

One must distinguish two kinds of identity of sense: on the one hand, there is the identity of sense of typographically similar sentences; this is simply *univocality*, the absence of ambiguity. On the other hand, there is the identity of sense of typographically different sentences; that is, *synonymy*. Because he neglected to make this distinction, Strawson attributed to the priority of sense over sign a power which, in fact, it does not have.

It is this first type of identity of sense which is *presupposed* by the fruitful use of logical schemata in argument. But it is the second whose existence one would have to prove in order to force the admission of propositional invariants; for the proposition which arouses the skepticism of nominalists is supposed to *transcend* differences between signs, as does *synonymy* in the usual sense, and as *univocality* does not.

Strawson's third argument is that the definition of *logical truth*, fundamental for logicians, which Quine formulated but without mentioning propositions, necessitates recourse to the contested notion. Indeed, Strawson maintains that when one *applies* Quine's definition to concrete cases formulated in natural language, one must *surrepticiously* appeal to the notion of univocality, that is, having a unique sense, and therefore at the same time to the notion of sense or, for short, intension. For example, one must have recourse to this in order to correctly classify the following as an empirical truth: 'If he was seen by the bank (in the sense of "river bank") then he was seen by the bank (in the sense of "commercial bank"), which a purely mechanical application of Quine's definition would have led us to rank among the logical truths. Very recently, however, Quine has answered this objection by giving a definition of univocality in *rigorous, extensional* terms. Thanks to the completeness result which obtains for first-order logic, it is possible to define syntactically the notion of 'logically demonstrable' in terms of a system of derivation. One could now say, suggests Quine, that a regimentation of our language is univocal if all the sentences which are logically demonstrable in that language are true.[17]

6. THE DEFINITION OF PROPOSITION IN TERMS OF THE PREMISES AND CONCLUSION OF AN INFERENCE

We began our investigation by attempting an axiomatic definition of proposition. This attempt ended in failure. To say that a proposition is something which satisfies the axioms of the propositional calculus is to expose ourselves, as has been shown, to the charge of circularity. Before looking in a radically different direction, one ought to ask oneself if it is not possible to formulate another definition of proposition and remain within the framework of logical syntax.

Ryle answered this question implicitly in 'If, So and Because' (1950); for he suggests in this paper that we carve out the concept of proposition or statement with the help of a characterization which appeals exclusively to the syntactic notions of 'premise' and 'conclusion'.

> Finally, it is an important, if not the important, feature of our use of words like "statement", "proposition", and "judgment", that any statement, proposition, or judgment can function as a premise or a conclusion in arguments. Suitability for what may be summarily called the "premissory job" is one of the main things that makes us reserve the title of "statement" for some sentences in distinction from all the rest.[18]

Ryle's suggestion calls for certain qualifications. It is actually appropriate to remark that it provides us with *necessary* but not *sufficient* conditions which must figure in any acceptable definition of the notion of proposition. Indeed, since there exists a logic of imperatives, it is difficult to deny that such imperative sentences can also serve as premise and conclusion in an inference, though they are *prima facie* not propositions.

Thus, one cannot take as a definition of *proposition* the assertion according to which a proposition is anything which can occupy the position of premise or the position of conclusion in an inference. However, one can see another merit in Ryle's suggestion: that of opening the way to formulating a criterion of *propositional identity*, a criterion which leads to a *definition by abstraction of* the notion of proposition in terms of inference: a proposition is the class of the mutually deducible sentences.

Although the definition by abstraction differs from the traditional definition, *operationally* it may be more useful than the latter. As Reichenbach remarked on another occasion, the notion of 'same weight as' is prior to that of 'weight' in contemporary science, because it correponds to an *effective procedure* of measurement. Actually, the scales indicate not the weight but the *equality of weight*. A definition by genus and specific difference would have been impracticable here. Thus, *mutatis mutandis*, it may be the case that it is easier to investigate *propositional identity* in discourse than *proposition*. Thus, one has reasons for not neglecting the resources which the criterion of identity of propositions offers to those who seek to define the concept of proposition. This question however, will be examined in a separate chapter.

REFERENCES

[1] A. Tarski, *Introduction to Logic*, Oxford University Press, 1941; 6th edn. 1954, p. 119.
[2] A. N. Whitehead and B. Russell, *Principia Mathematica* to *56, Cambridge University Press, 1962, p. 402–403.
[3] W. V. O. Quine, 'Ontological Remarks on the Propositional Calculus' (1934) in *The Ways of Paradox*, Random House, New York, 1966, p. 61.
[4] Quine, *Ibid.*, p. 62.
[5] W. V. O. Quine, 'Logic and the Reification of Universals', *From A Logical Point of View*, 1953; Harper Torchbooks, New York, 2nd edn., 1961, p. 109.
[6] Quine, *Ibid.*, p. 111.
[7] G. G. Granger, *Wittgenstein*, Seghers, Paris, 1969, p. 40.
[8] See G. E. Moore, 'Wittgenstein's Lectures' (1954), *Philosophical Papers*, Allen and Unwin, London, 1959, p. 296.
[9] See I. Copi, *Symbolic Logic*, Colliers, Macmillan, London, 3d edn. 1967, p. 138.
[10] R. Wells, 'Meaning and Use', *Word* X, (1954) 244.
[11] P. F. Strawson, 'Propositions, Concepts and Logical Truths', *Philosophical Quarterly* 7, (1957) 23.
[12] P. F. Strawson, 'Paradoxes, Posits, Propositions', *Philosophical Review* LLXXVI, (1967) 214–219.
[13] Strawson, 'Propositions, Concepts and Logical Truths', *op. cit.*, 19.
[14] Strawson, *Ibid.*, 20.
[15] W. V. O. Quine, *Word and Object*, Wiley, New York, 1960, p. 65n.
[16] P. F. Strawson, 'Paradoxes, Posits, Propositions', 216–217.
[17] W. V. O. Quine, 'Replies', *Synthese* 19 (1968), 296; repr. in *Words and Objections*, D. Davidson and J. Hintikka (eds.), Reidel, Dordrecht, 1969.
[18] G. Ryle, 'If, So, and Because' in M. Black (ed.), *Philosophical Analysis*, Cornell, Ithaca, 1950, p. 325.

CHAPTER III

A SEMANTIC DEFINITION OF PROPOSITION IN TERMS OF TRUTH AND FALSITY

1. THE ARISTOTELIAN DEFINITION OF PROPOSITION IN TERMS OF TRUTH

In *De Interpretatione* Aristotle states a *necessary condition* for the application of the word 'proposition'. He writes: "We call propositions those only that have truth or falsity in them. A prayer is, for instance, a sentence but neither has truth nor has falsity".[1]

Is not this necessary condition at the same time a *sufficient* condition which would allow us to use it in a definition of the concept of a proposition? This question calls for a varied answer. At first blush, the answer must be negative; for propositions are indeed not *the only* objects for which one may predicate truth and falsity. One attributes these predicates also to sentences, statements, beliefs and judgments. In any case, if it were possible to prove that truth and falsity are applied *primarily* to propositions and only apply to others terms of the list in a derivative manner, one could answer positively and maintain that the property of being true or false gives us a definition of the proposition.

The thesis according to which the *proposition* as distinct from the sentence is, above all, the 'vehicle' for truth and falsity is an opinion that prevailed for a long time. Since we are going to oppose it, it is important to recall precisely the classical way of formulating it, and the traditional arguments involved in its defense.

In his great treatise on logic (*Logic*, 1921), W. E. Johnson declares that "A proposition is that of which truth and falsity can be significantly predicated".[2] He neatly distinguishes the proposition from the sentence:

> It has been generally held that the proposition is the *verbal expression* of the judgment; this, however, seems to be an error, because such characterisations as true or false cannot be predicated of a mere verbal expression, for which appropriate adjectives would be 'obscure', 'ungrammatical', 'ambiguous', etc.[3]

Likewise, in *General Logic* (1931) R. Eaton joins Johnson in affirming that

> The proposition must be distinguished from the *sentence*, the combination for words or signs through which it is expressed; from the *fact*, ... and from the *judgment* ... . These distinctions can be made by noting that certain adjectives, true, false, doubtful,

impossible, and others, apply to propositions but do not apply in the same sense, if at all, to sentences, facts, and judgments.

When we say, "It is true (or doubtful) that *Joffre won a decisive victory at the first battle of the Marne*", we are not calling attention by the adjectives true or doubtful to anything about the sentence printed in italics, neither to its grammatical form nor its component words. We could correctly say that this is an English sentence, but the proposition is not English; it could be equally well expressed in Turkish.[4]

The notion of a proposition as distinct from a sentence thus seems to answer a precise need, namely, that of transcending linguistic particularities, which makes sentences inappropriate vehicles for truth. In 'Truths of Logic' (1945–1946), Kneale writes:

When we refer to a certain proposition of Euclid we do not mean a certain sentence in Greek, and so too the older logicians when they spoke of propositions seem to have been abstracting from the peculiarities of any particular language. No one supposes that Aristotle's logic was Greek in a sense in which Euclid's geometry was not.[5]

Thus, the utility of the concept of proposition here seems to be well established: the definition of the proposition in terms of truth confers on it an extra-linguistic or translinguistic status which protected it from nominalistic enterprises of a Hobbes, who asserted that "truth is an attribute of parole, not of things". But Tarski's work on *The Concept of Truth in Formalized Languages* (1933), put everything into question again and furnished, as we shall see, new arguments for the nominalists.

2. THE INFLUENCE OF THE SEMANTIC DEFINITION OF TRUTH ON THE CONCEPT OF PROPOSITION

Tarski has demonstrated that one cannot define the expression 'true sentence' in a satisfactory way (that is, in a way which avoids the Paradox of the Liar) without distinguishing the object-language from the metalanguage. The former contains sentences to which the predicate 'true' is to be applied; the latter contains this predicate among others. It follows that, if one wishes to define 'true' for sentences of the metalanguage, one must carry out the definition in a metametalanguage.

A corollary which seems to spring immediately from these exigencies for which Tarski revealed the basis, is that the concept of 'truth' is not an *absolute* but a *relative* concept. 'Truth' is, from now on, relative to a specified level of language.

No doubt in order to rebut the assertions, of the philosophers Johnson, Eaton and Kneale, who maintain that the predicate 'true' *transcends* particular languages and that it must be predicated of propositions rather than sentences, one ought to be able to show that the predicate 'true' is not only relative to a *level* of language, but also to a *particular language*.

The transition between these two theses is nevertheless easy to justify. Levels of language are indeed always levels of one definite language. In the case of formal languages for which the predicate 'true' is defined in Tarski's way, the class of well-formed formulas belonging to the language must have been specified rigorously and exhaustively by means of recursive rules. For natural languages the problem is not very different. Although the formation rules in this case codify a pre-existing structure, nothing authorizes us to say that this structure is the same for all languages, nothing authorizes us to postulate some kind of *Ursprache*. Besides, as we shall see later, in the case of natural languages, linguistic 'stratification' is also introduced from the outside, thanks to appropriate conventions.

If we admit, as Tarski's results suggest, that the predicate 'true' is incomplete as long as one has not specified the language to which it relates ('true in L'), we are obliged to recognize that it is predicated of sentences rather than propositions. However, it does not follow that it is assimilable to the adjectives 'obscure', 'ungrammatical', 'ambiguous', as Johnson fears. Actually, the predicate 'true' has, in this respect, an interesting property to which Quine has drawn attention in *Philosophy of Logic* (1971).

Consider Tarski's famous paradigm: "Snow is white" is true if and only if snow is white. Take the word 'true'. The *reference* of this predicate, that is, the object to which it is attributed, is here a sentence. Its *grammatical subject* is a quoted sentence, that is, a name of a sentence, formed by putting the sentence inside quotation marks. The presence of quotation marks *divests* the words 'snow' and 'white' of their objective meaning. They no longer refer to a substance and a colour, but to words. And it is precisely the use of the predicate 'true' which reinstates the extralinguistic content of the sentence that putting it in quotation marks had removed in the first place; it *annuls* the effect of quotation. The escalation achieved by quotation marks in attributes of truth is only temporary. Quine writes that "The truth predicate is a reminder that, despite a technical ascent to talk of sentences, our eye is on the world".[6] In other words, confirming the truth of the sentence 'snow is white' is at the same time confirming the whiteness of snow, though it is not solely that. If one follows Quine, one would subscribe to the thesis according to which predicating truth transcends the *sentence*, not, of course, to arrive at the *proposition*, but to arrive at *reality*. It is, therefore, possible to do justice to Johnson's and Eaton's intuitions without recourse to propositions.

With respect to the preoccupation with transcending particular languages to formulate universal logical truths, as echoed by Kneale, it is possible to do him justice by using schematic letters admitting sentences as substitutes.

But to return to Tarski, one might really question whether his results actually have the philosophical impact that we have attributed to them. His investigation, of course, bears on the definition of truth for formalized languages; but do natural languages have the same features? Is natural language, for example, hierarchical as a formalized language must be if it is to be free of paradox? Tarski and several authors after him have asked themselves that question.

Tarski recognizes in colloquial language a universality and homogeneity which excludes differences of level and partitions, and this separates such languages from formalized systems:

> It would not be in harmony with the spirit of this language if in some other language a word occurred which could not be translated into it; it could be claimed that 'if we can speak meaningfully about anything at all, we can also speak about it in colloquial language'.[7]

However, he does not conclude, like certain others, that colloquial language is incoherent. Rather, he believes that the notion of 'sentence' in natural language is not sufficiently clear for the question of consistency or inconsistency to be decided. He writes:

> ... this language is not something finished, closed, or bounded by clear limits. It is not laid down what words can be added to this language and thus in a certain sense already belong to it potentially. We are not able to specify structurally those expressions of the language which we call sentences, still less can we distinguish among them the true ones.[8]

The facts seem to agree with the description given by Tarski. Indeed, the notion of sentence in natural language is not clear. The distinction between declarative sentences – the only ones that can be true or false – and non-declarative sentences cannot be drawn if one considers only the *form* of the sentences concerned.

As Ch. Perelman notes in 'Logique, Language et Communication' (1958), 'formal logic cannot by itself decide when one is confronted by an assertion'.[9] He gives as an example the sentence "You shall not kill", which, *according to context* may be classified as indicative or imperative. If *context* is required, it is clear that it is because the *form* of the sentence is not enough to decide its status.

These considerations about the fluidity of natural language go together with what we said in the preceding chapter about the form of reasoning in natural languages. However, one would go too far in concluding that Tarski's theory simply does not concern natural language. The position which appears more satisfying in this respect is Popper's, who writes in *Conjectures and Refutations* (1963):

> The view that his theory is applicable only to formalized languages is, I think, mistaken. It is applicable to any consistent – more or less – 'natural' language. So we must try to learn from Tarski's analysis how to dodge its inconsistencies; which means, admittedly, the introduction of a certain amount of 'artificiality' – or caution – into its use.[10]

If this is so, certain arguments formulated in (standardized) natural or semi-natural language can fully claim *the same* rigor as arguments formulated in symbolic language. This is contrary to what Kemeny claims, who writes that "ordinary English is a language not suitable for logical arguments".[11] But, in *compensation*, natural language is exposed to the antinomy of the liar; for, by virtue of its homogeneity and universality, it places expressions of the object language and those of the metalanguage on the same level.

Thus, one must fend off this paradox, which is what Bar-Hillel set out to do in 'Do Natural Languages Contain Paradoxes?' (1966).[12] According to him, liability to paradox is the foremost property of natural languages which must be removed by means of a distinction between sentence and statement.

3. USE OF THE DISTINCTION BETWEEN SENTENCE AND STATEMENT AS A SOLUTION TO THE PARADOX OF THE LIAR IN NATURAL LANGUAGE

The sentence 'this sentence is false' falls outside the truth-false dichotomy. As is well known, we are caught in the trap of a paradox when we try to ascribe to it either truth or falsity. Indeed, it can be easily shown that it is true *if* and *only* if it is false. Thus one cannot say that it is true *or* false. This paradoxical sentence therefore seems to constitute a counter-example to the nominalist who wishes to attribute alethic bivalence to sentences, i.e. the capacity of being true or false, which is usually invoked in order to characterize the proposition.

Bar-Hillel has established a distinction between sentence and statement which the nominalist could use to remove this counter-example. This distinction is by no means an *ad hoc* one. One needs it in order to give an account of performatives. As Austin has shown, to utter, under certain conditions, a declarative sentence such as 'From now on this square shall be called Place Général-de-Gaulle' is not to make a statement, *but is to accomplish an action by means of language*. On the other hand, a declarative sentence containing an 'indexical expression' such as 'I am now hungry' has no permanent truth value, as the sentence-type changes its truth value with context. This same sentence uttered by two different persons serves to make two different statements, whereas the two sentences 'You are hungry' and 'I am hungry' may be used to make one and the same statement.

According to Bar-Hillel, it is not declarative sentences characterized by their grammatical form which must be considered subjects of truth and falsity but, rather, statements. These have neither logical nor grammatical form; and as Ferdinand Brunot already pointed out in *La Pensée et la Language*, they may be expressed by means of declarative sentences of different forms. They are for this reason incapable of functioning, like sentences or propositions, as terms in logical relations. Bar-Hillel is aware of this gap which seems, after all, rather serious. He writes: "Notice that the transition from statements to sentences is absolutely necessary for certain purposes such as the application of formal logic, ...".[12]

Bar-Hillel's originality lies in the way he employs the notion of statement to avoid the Paradox of the Liar without stratifying language into levels, nor admitting with the opponents of nominalism either that *some* sentences are not alethically bivalent or that propositions alone possess this property. Instead of admitting that the isolated sentence 'This sentence is false' is neither true nor false – contrary to the case with the majority of declarative sentences – Bar-Hillel prefers to say that it does not succeed in expressing a statement, that is, an entity susceptible of being true or false, and that logical reasons prevent this.

If one does not seek to construct a unified theory which accounts both for the syntactic and the semantic structure of the sentence, Bar-Hillel's solution is tempting; but it has an unsuspected doctrinal implication which we must examine carefully. It revives the controversy between realists and nominalists concerning the ontological status of propositions. *This arises from the fact that nominalists reject the notion of statement almost as much as the notion of proposition.*

Thus Quine, for example, in 'Mr. Strawson on Logical Theory' (1953), reproaches the British logician for the distinction between sentence and statement, which Strawson draws for other ends than Bar-Hillel. Quine writes: "The risk is that of hypostatizing obscure entities, akin perhaps to "propositions" or "meanings" or "facts" or "states of affairs", and reading into them an explanatory value which is not there".[13] This is also Lemmon's opinion, writing in 'Sentences, Statements and Propositions' (1967) that

... the ontological status of statements and propositions is peculiar. For they are certainly not linguistic entities, as sentences are — do not belong to a language — nor are they spatio-temporal particulars, locatable at a position in space–time, like physical objects or even events and processes. Thus, there is a *prima facie* case against postulating them, if only in accordance with Ockham's razor, unless we have to.[14]

Mates is even more categorical. In *Elementary Logic* (1965) he affirms that statements have "a rather serious drawback, which, to put it in the most severe manner is this: they do not exist".[15]

Before ratifying Bar-Hillel's contribution we must answer the following two questions:

(a) Are there reasons *other* than those he gives for introducing the distinction between statements and sentences?

(b) Is it possible, thanks to an appropriate modification, to keep the insight of his argument without abandoning the view that sentences rather than statements are above all the subjects of the predicates 'true' and 'false'?

4. THE ONTOLOGICAL STATUS OF THE DISTINCTION BETWEEN STATEMENTS AND SENTENCES

The distinction between sentences and statements has been outlined repeatedly. We find it in Austin's 'Truth' (1950); in Strawson's *Introduction to Logical Theory* (1952); and, finally, in Bar-Hillel's 'Indexical Expressions' (1954). But only recently has the question of its *ontological* scope been considered in depth. If the problem of the ontological status of statements is a manifestation of the problem of the ontological status of propositions, it is essential for us to come to terms with this question. We are thus obliged to arbitrate between Cartwright's and Stroll's arguments raised by this issue.

In 'Propositions' (1962) and 'Propositions Again' (1968), Cartwright defends a realist conception of the statement. His argument is substantially the following:

Consider two persons, A and B, who both utter the sentence 'I am tired'. One sees immediately that there are two specimens of *the same sentence*, but *different statements*. Spoken by A, the sentence describes a fact which relates to the state of A's neuro-psychical system; spoken by B it describes a fact which relates to B's state. One may, therefore, admit the following as true premises:

(1) The sentence uttered by A is identical with the sentence uttered by B.
(2) The statement made by A is not identical with the statement made by B.

From this one would want to deduce the conclusion:

(C)   The sentence uttered by A is not identical with the statement made by A.

Cartwright thought that the above argument requires an additional premise in order to be valid; it can be formulated as follows:

(3)   If the sentence uttered by A is identical to the statement made by A, then the sentence uttered by B is identical to the statement made by B.

The argument which he gives, appears to establish in an irrefutable manner the basis for the distinction between statements and sentences by showing that this *verbal distinction* corresponds to an *actual difference*. But this does not yet suffice, I believe, to show the *autonomous existence* of statements, the *independence* of statements with respect to sentences. It is this that would have been necessary to take up Mates' challenge.

Stroll goes still further in his reservations about Cartwright's conclusions. He is not content in affirming, as we have done, that the difference between statements and sentences does not *suffice* to assure their mutual independence; he maintains that those who claim the contrary are undoubtedly begging the question. That is the sense of the following passage from 'Statements and Propositions: the Detachment Argument' (1970):

Despite Cartwright's asseverations to the contrary, it does not follow from the fact that two different linguistic expressions may be used in making the same statement that the statement they make cannot be identical with both of them.

... whether the above inference is valid depends in part, upon the assumption that statements are entities in their own right. On that assumption, it does indeed follow that a given *thing* cannot be identical with both of two other things which differ from one another. But if the assumption that statements are *things* ... is not made then there is no reason to accept the argument as valid. If statements are not entities, there is no *thing* which fails to be identical with both of the differing locutions in question.[16]

This ingenious argument of Stroll's is not valid. If one considers it carefully, one notices that it presupposes that *entities* or *things alone* can be identical or different; in other words, he supposes that inference rules from the logic of identity only apply to names which denote entities or to variables which take things as values. This is wrong. One can very well construct an *ontologically neutral* logic of identity, the variables of which receive a 'substitutional' rather than a referential interpretation; that is, a logic whose variables are halfway between usual variables and Quine's schematic letters.

The inference rules for identity used by Cartwright would remain intact

A SEMANTIC DEFINITION OF PROPOSITION

in this logic which does not take a stand on the status of terms linked by the relation of identity (which does not necessarily make them 'entities'). In other words, the validity of Cartwright's argument is not subordinate to the prior adoption of a postulate according to which statements are *entities, things*. We conclude that Cartwright did not beg the question as Stroll charges.

N.B. It is easy to prove, by *reductio*, the unconditional validity of Cartwright's reasoning.

Let us call

$P(A)$ and $P(B)$ the sentences uttered by A and B,
$E(A)$ and $E(B)$ the statements made by A and B.

Cartwright's three premises are:

(1) $P(A) = P(B)$
(2) $E(A) \neq E(B)$
(3) $P(A) = E(A) \cdot \supset \cdot P(B) = E(B)$.

The conclusion is:

(4a) $P(A) \neq E(A)$.

Suppose that:

(4b) $P(A) = E(A)$;

we obtain:

(5) $P(B) = E(B)$ by virtue of (3) and (4b)
(6) $P(A) = E(B)$ by virtue of (5) and (1)
(7) $E(A) = E(B)$ by virtue of (6) and (4b).

Now:

(7) contradicts (2).

Stroll likewise considers analogies: he rejects the *realist* conception of the statement according to which the latter is separable from the sentence in the way that *freight* is separable from the *vehicle* transporting it. He attempts to make us accept a *conceptualist* conception, according to which a statement is *inseparable* from, and even *identical* with, a sentence in certain respects, though *different* in others, like the action of *taking* can at once be and not be the *same* as the action of *stealing*. According to Stroll, the second analogy is the one to be retained. The reason justifying this preference is that one cannot give a description of a *statement* which is independent from a description of the sentence expressing it; whereas one can perfectly well give a description of the *freight* of a vehicle which is independent of a description of the *vehicle* carrying it.

Cartwright rightly rejects this argument by Stroll. The same argument would actually allow us *a pari* to draw false conclusions, namely to prove that the number 11 has no existence independent of the numerals since one cannot designate it without using a numeral (unless one uses Russell's logistic definitions). This objection against Stroll appears decisive to us. One knows, indeed, that the *existence* of numbers does not depend upon the operation of counting even if the act of *naming* a particular number is subordinate to the use of a numeral. Cantor's diagonal argument definitely establishes this result; Stroll's argument would oblige us to deny it.

But if Stroll's inference is not valid, his *premise* is nevertheless true and interesting: one cannot *have access to* a statement except *by way of* the sentence. Thus, we must ask ourselves if there are no other interesting conclusions one might infer, validly, from this premise. There is one, in fact; and it alone justifies the importance we attached to the polemics between Stroll and Cartwright. More particularly, one may exploit the observation which served as the premise for Stroll's incorrect inference to demonstrate, this time correctly, that *sentences* rather than *statements* are the *primary* vehicles of truth and falsity.

This thesis, which at first blush contradicts Bar-Hillel's, is of the greatest interest to us. If it is true, we then have the right to extend to natural language the practice (established by Tarski for formalized languages) of attributing truth and falsity to *sentences*. Robin and Susan Haack in 'Token Sentences, Translation and Truth-Value' (1970)[17] constructed original arguments precisely in support of the thesis, according to which truth and falsity apply to sentences *prior to* applying to statements. We must now examine these arguments.

5. TRUTH AND FALSITY APPLY TO SENTENCES BEFORE APPLYING TO STATEMENTS

That statements, though different from sentences, are *tributary* to sentences is a truth upon which we do not have to insist. The statement is, by definition, that which has been stated, that is, the result of an utterance. It cannot exist prior to the speech act.

The statement cannot exist in the mind of a speaker in the manner of an intention any more than a promise is an intention to promise. This follows clearly from Austin's work, the importance of which for our problem will be shown in the next chapter.

The originality of Robin and Susan Haack's contribution consists in having shown that from the *dependence* of statements on *sentences* one may conclude that the predicates 'true' and 'false' apply primitively to sentence-tokens, and derivatively to statements. They, themselves, have clearly laid out the scope of their argument. It is, however, useful to put it into concrete form by means of an example.

Consider two *occurrences* of the sentence 'It is cold here'. Before settling the question whether these two sentences correspond to the *same* statement or to *two different statements*, we have to know whether they have the same truth-conditions. In other words, it is not the statement itself which is subsidiary to the sentence; rather, the identification of the statement by reference to its truth-conditions depends upon the prior identification of the sentences which express it. The dyadic predicate 'x has the same truth-conditions as y' therefore applies to sentences *before* applying to statements. Obviously, one may conclude that this is also valid for the monadic predicate 'x is true'.

Does this mean that Bar-Hillel's claim is false and that statements are not the primary bearers of truth and falsity? If this were the case, we ought to reject the skillful parry with which Bar-Hillel obstructed the Paradox of the Liar in natural language. Fortunately, we can avoid this conclusion and reconcile the two opposed theses. To this end it suffices to take cognizance of a judicious remark by Robin and Susan Haack at the beginning of their study: namely, that the thesis according to which sentences are the primary bearers of truth and falsity by no means implies that *all* sentences are bearers of truth and falsity.

In other words, one may very well remove from *certain* sentences their alethic prerogatives, as one takes civil rights away from certain persons, and may thus dispel threats of inconsistency directed at natural language. In short, one may retain Bar-Hillel's doctrine while dispensing with his terminology.

6. THE SEMANTIC THEORY OF TRUTH AND THE CORRESPONDENCE BETWEEN LANGUAGE AND REALITY

The semantic theory of truth frees the nominalist from the obligation of accrediting *propositions* or *statements* as the 'vehicles' of truth and falsity.

But its utility does not end here. This theory also frees the nominalist from the obligation of assuming another kind of obscure metaphysical entity: *facts*. Finally, it contains, as we shall see, very important pieces of the framework of the general theory which we are trying to construct.

The most elaborate theory of truth does not present truth as an *adequatio rei et intellectu*, but rather as a correspondence between propositions and facts. One says that a false proposition does not correspond to anything. Philosophers who have professed one or another form of logical atomism (Russell, Wittgenstein) conferred on certain facts the status of independent logical entities ('atomic facts').

According to this theory — overlooking the differences between Russell's and Wittgenstein's logical atomism — what makes the sentence 'Socrates is bald' true, is not the bald Socrates, that is, an *individual* having a certain property, but the *fact* that Socrates is bald. We shall demonstrate later on, when treating the ontological aspects of the proposition systematically, that the concept of a fact is not an admissible one, and that there are compelling reasons for dispensing with it.

Renouncing facts is not necessarily to renounce a realist conception of truth and to take the dead-end routes of idealism or pragmatism. Without facts we can still explain the difference between the truth and falsity of sentences, but in another way: for example, by saying that this difference has its root in the *objective reality*. One must, recognize however, that such an explanation is not as subtle as the preceding one. To say that true sentences are those which correspond to reality is to give an account of the truth of sentences which is *uniformly* the same for all. The explanation of truth in terms of sentences corresponding to facts, on the contrary, associates distinct facts with distinct true propositions.

One of the merits of Tarski's theory of truth is to provide a solution that is entirely in accord with the concept of truth as a correspondence between propositions and facts. The semantic theory of truth indeed contains an element of correspondence. This has been known for a long time, but not always correctly understood.

One sometimes locates the realist element in Tarski's theory in the material condition of adequacy which he imposes as *preliminary* to any definition of truth. This condition of adequacy is stated thus: 'S' is true if and only if S. Admittedly, this material condition of adequacy reveals that although truth is predicated of linguistic entities (sentences), 'our eye is on the world'. As Quine puts it so well, by calling the sentence 'snow is white' true, we call snow white: "The truth predicate is a device of disquotation".[18]

Moreover, the material condition says that for the sentence in brackets to be true it is *necessary and sufficient* that snow is white, from which it follows that it is not necessary at all that people exist to observe it. The *meaning* of the sentence is perhaps subordinate to the existence of people, but *truth* is not directly so subordinate. Thus Tarski's concept of truth, as it emerges from the material condition of adequacy, agrees with Aristotle's view expressed in his treatise *De anima*:

... truth and falsehood [are] in the same class with the good and the evil. Yet in this, at any rate, they differ, that the former are absolute, the latter relative to some one concerned.[19]

Nevertheless, it is not in the material condition of adequacy that one finds the interesting element of *correspondence* which is present in Tarski's truth theory.

Davidson correctly identifies this element of *correspondence* in 'True to the Facts' (1969), where he shows that "The semantic conception of truth as developed by Tarski deserves to be called a correspondence theory because of the part played by the concept of satisfaction".[20] Tarski himself gives a very concise resume of his theory of satisfaction in 'The Semantic Conception of Truth' (1944):

To obtain a definition of satisfaction we have rather to apply again a recursive procedure. We indicate which objects satisfy the simplest sentential functions; and then we state the conditions under which given objects satisfy a compound function ... .

Once the general definition of satisfaction is obtained, we notice that it applies automatically also to those special sentential functions which contain no free variables, i.e., to sentences. ... Hence we arrive at a definition of truth and falsehood simply by saying that *a sentence is true if it is satisfied by all objects, and false otherwise.*[21]

What escaped the sagacity of earlier commentators but not Davidson, is that for Tarski *satisfaction* plays the rôle that in traditional theories fell to *correspondence*. It is satisfaction which actually establishes a relation between reality and language.

In the case of normal propositional functions, that is, in the case of propositional functions admitting individual variables, the entities capable of satisfying these functions are simply *objects*. In this case the objective correlate is, therefore, neither a *fact* nor *reality* taken as a whole, but one or several individuals. But when we come to sentences, that is, to propositional functions *without* free variables, the correlate ('the satisfier') is reality in its entirety. This is what Davidson intends when he writes:

All true sentences end up in the same place, but there are different stories about how they got there; a semantic theory of truth tells the story for a particular sentences by running through the steps of the recursive account of satisfaction appropriate to the sentence.[22]

Tarski's idea as made explicit by Davidson is therefore that a true sentence is true *for* reality as a whole (is satisfied by all objects), but it is also constructed with the help of propositional functions which are true of *individuals* (ordered pairs, triples, etc., of objects) which satisfy it. Quine formulates the same thesis in succinct terms: "The definition of *truth* in terms of satisfaction is easy indeed: *satisfaction by all sequences*".[23] Therefore, one no longer needs to invoke entities *sui generis*, facts, to serve as the second term in the relation of correspondence, which, according to the supporters of the correspondence theory of truth, exists between true sentences and reality. We take instead as the second term of the correspondence the set of all sequences, that is to say, the entire domain of individuals and *n*-tuples of individuals. The correlate of a false sentence is the empty set.

This is of *great importance* for the *general theory* we are trying to construct. Indeed, we shall see that the solution put forward by Davidson to explain how sentences *identical* in truth value can have different correlates agrees perfectly well, and even partly coïncides, with those that we shall present to solve the problems posed by the *creativity of language* and the *capacity of false sentences to signify*.

REFERENCES

[1] Aristotle, *Organon; On Interpretation*, 17a (1–5), Loeb Classical Library, Harvard University Press, Cambridge, and Heinemann, London, 1955, p. 121.

[2] W. E. Johnson, *Logic*, (1921) Vol. I; repr. Dover, New York, p. 1.

[3] Johnson, *Ibid.*, p. 1.

[4] R. Eaton, *General Logic: an Introductory Survey*, Scribner's Sons, New York, 1931; new edn. 1959, p. 12.

[5] W. Kneale, 'Truths of Logic', *Proceedings of the Aristotelian Society* 46, (1945–1946) 208.

[6] W. V. O. Quine, *Philosophy of Logic*, Prentice-Hall, Englewood Cliffs, N.J., 1970, p. 12.

[7] A. Tarski, *The Concept of Truth in Formalized Languages*, trans. by J. H. Woodger, Clarendon Press, Oxford, 1956, p. 164.

[8] Tarski, *Ibid.*, p. 164.

[9] Ch. Perelman, 'Logique, Language et Communication', *Actes du XIIIe Congrès International de Philosophie*, Rapports introductifs, Sansoni, Firenze, 1958, p. 131.

[10] K. R. Popper, *Conjectures and Refutations*, Routledge and Kegan Paul, London, rev. edn., 1976, p. 223.

[11] J. C. Kemeny, 'Semantics as a Branch of Logic', *Encyclopedia Britannica*, Vol. 20 (1957) p. 311.

[12] Y. Bar-Hillel, 'Do Natural Languages contain Paradoxes?', *Studium Generale* (1966) 395; repr. in Bar-Hillel, *Aspects of Language*, North Holland, Amsterdam, 1970, p. 282.
[13] W. V. O. Quine, 'Mr. Strawson on Logical Theory', *Mind* (1953); repr. in *The Ways of Paradox*, Random House, New York, 1966, p. 143.
[14] E. J. Lemmon, 'Sentences, Statements and Propositions', in P. Williams and A. Montefiore (eds.), *British Analytical Philosophy*, Routledge and Kegan Paul, London, 1966, p. 100.
[15] B. Mates, *Elementary Logic*, Oxford University Press, 1965, p. 8.
[16] A. Stroll, 'Statements and Propositions: The Detachment Argument', *International Logic Review* (1970) 95.
[17] R. Haack and S. Haack, 'Token-Sentences, Translation and Truth-Value', *Mind* 79, (1970) 51–52.
[18] W. V. O. Quine, *Philosophy of Logic*, p. 12.
[19] Aristotle, *De Anima* 431b (10), Hicks (ed.), Hakkert, Amsterdam, 1965, p. 143.
[20] D. Davidson, 'True to the Facts', *Journal of Philosophy*, (1969) 758.
[21] A. Tarski, 'The Semantic Concept of Truth', *Philosophy and Phenomenological Research*, 4 (1944); repr. in H. Feigl and P. Sellars (eds.), *Readings in Philosophical Analysis*, Appelton-Century-Crofts, New York, 1949, p. 63.
[22] Davidson, *Op. cit.*, p. 759.
[23] W. V. O. Quine, *Philosophy of Logic*, 1970, p. 38.

CHAPTER IV

THE PRAGMATIC DEFINITION OF PROPOSITION IN TERMS OF ASSERTION OR ASSERTABILITY

1. THE PRAGMATIC DEFINITION OF PROPOSITION IN TERMS OF ASSERTABILITY

We saw that Johnson defined the proposition as that of which one can predicate the 'true' and the 'false'. But, this distinction has its roots in another, according to that author, the distinction between the correct and erroneous:

> Thus, though we may predicate of a certain proposition ... that it is true or that it is false, what this ultimately means is, that any and every thinker who might at any time assert the proposition would be either exempt or not exempt from error.[1]

As the distinction between correct and mistaken appears fundamental to him, Johnson constructs a new definition of the proposition whose role is to bring out the relations between the proposition and the assertion.

> In order to mark the important distinction, and at the same time the close connection, between the proposition and the act of assertion, I propose to take the term 'assertum' as a synonym for 'proposition'... Thus, the assertum will coïncide, not exactly with that which *has been* asserted, but with that which is, in its nature, assertible.[2]

One may find it surprising that Johnson considers the opposition between 'correct' and 'mistaken' as *more fundamental* than the opposition of 'true' and 'false'. The notion of error does not actually seem to deserve this privilege. It is not more fundamental, but simply more general and less precise, than the notions of 'true' and 'false' in the sense that it applies likewise to cases of linguistic error (the subject was wrong about words) and to factual error (the subject was mistaken about 'facts'), whereas the opposition of 'true' and 'false' applies only to the second case. Moreover, the notion of error is an *epistemological* concept which the notion of falsity is not. One must, however, give Johnson credit for having drawn attention, by shifting toward pragmatics, to the *problematic* character of the false, and to the asymmetry of the false with respect to the true, an asymmetry for which we shall have to account.

From his pragmatic definition of the notion of proposition, Johnson draws certain conclusions with respect to its ontological status. According to him, "the proposition is not so to speak a self-subsistent entity, but only a factor in the concrete act of judgment".[3] In other words, the proposition as conceived by Johnson cannot exist in an autonomous way like Bolzano's *Satz an sich*.

One readily understands that the *statement* is subsidiary to the *utterance*. By definition, the statement is 'that which has been stated'. But the same does not apply for the proposition. Why could not the proposition, understood as that which is assertable, exist prior to the assertion, as a flammable object may exist prior to being on fire?

The statement is the *result* of the act of stating, but the proposition could be conceived as the *object* of the act of stating. The contrast object-result is exemplified by the following sentences

(1)   He is reading a book
(2)   He is writing a book

which I borrow from Lyons together with the following comment: "In (1) the book that is referred to exists prior to, and independently of, its being read; but the book referred to in (2) is not yet in existence — it is brought into existence by the completion of the activity described by the sentence".[4]

The Saussurian opposition between language and speech has a bearing on the issue and could inspire a new line of defense of Johnson's thesis. One might argue, as Benveniste and Ryle did, that words and constructions belong to *language* and exist prior to their use, whereas sentence is the unit of *speech*. If so one could say that the meaning of a sentence, i.e. the proposition, is brought into existence by the uttering of the sentence.

This line of defense of Johnson's thesis is not convincing. It is the *utterance* of the sentence rather than the *sentence* itself which should be treated as the unit of speech. The infinite and recursive set of well-formed sentences can be said *virtually* to *exist* in the formation rules. If a recursive semantics which maps well formed sentences onto meanings, i.e. onto propositions, is added to the recursive syntax, it seems reasonable to assume a recursive set of propositions existing prior to statements. Modern developments in *formal semantics* seem to run counter to Johnson's claim.

There are other definitions of proposition beside Johnson's which broadly speaking, relate to pragmatics by the reference they make to the speaker or to the act of propounding. The one which I will consider next is Geach's.

2. THE DISTINCTION BETWEEN PROPOSITION AND STATEMENT FROM A PRAGMATIC PERSPECTIVE

In *Assertion* (1965), Geach defines the proposition as follows:

> When I use the term "proposition", as I did just now, I mean a form of words in which something is propounded, put forward for consideration; it is surely clear that what is then put forward neither is *ipso facto* asserted nor gets altered in content by being asserted.[5]

Geach, inspired by Frege, thus does not restrict himself to saying, like Johnson, that the proposition is that which is assertable. He adds that it is *at times asserted, at times simply considered*, and that it remains the *same proposition* in either case. This is an extremely important trait of propositions and one which we think is sufficient to defeat the assimilation of the sentence to a tool.

This latter analogy indeed masks an important difference. Whereas a tool may be *used* or *exhibited in a showcase*, a proposition may be *asserted, proposed* or *quoted*, quotation being for the sentence, the equivalent of exhibition for the tool. The use of the proposition, thus, is more complex; it has a "greater logical multiplicity" than the use of a tool.

If one considers this, one will understand why it is unacceptable to conflate *propositions* with *statements*. Indeed, by definition the statement is always asserted. As Geach says, "The notion of an unasserted statement may appear a contradiction in terms".[6] Now, we need the notion of an unasserted *something*, i.e. of a proposition, if we want to account for the modus ponens, since one of the premises of this argument is a hypothetical statement of the form "If p then q".

Ryle tried to refute this objection in advance. In 'If, So and Because' (1950), he argues that in practice, we do not go from the statements 'p' and 'if p then q', where the second occurrence of 'p' and the occurrence of 'q' are unasserted propositions, to the statement 'ergo q', but that we go directly from the asserted proposition 'p' to the asserted proposition 'q' in a straightforward inference 'p therefore q' where 'p' and 'q' are both asserted, i.e. are statements.[7] It is only afterwards that we extract by abstraction hypothetical statements of the form 'If p then q'. Logic resembles culinary practice, where the preparation of a dish precedes the codification of this practice in the form of recipes.

Ryles' argument, however, is not conclusive. Even if *genetically* we start without using hypothetical statements containing unasserted propositions, we definitely do use such hypothetical statements when we codify inferences.

The sort of proposition which is needed to account for the distinction between 'asserted p' and 'unasserted p', is, for that matter, completely immune to criticism from a nominalistic point of view. We could just as well speak of 'asserted *sentence*' versus 'unasserted *sentence*'.

Still another concept of proposition emerged from Speech Act Theory. Searle introduced proposition as an essential part of the illocutionary act. The general form of illocutionary acts is $F(p)$ "where the variable 'F' takes illocutionary force indicating devices as values and 'p' takes expressions for propositions".[8] Before examining this new concept of proposition, a few words have to be said about the key notions of illocutionary act and illocutionary force.

3. AUSTIN'S DISTINCTION BETWEEN LOCUTIONARY AND ILLOCUTIONARY ACTS

Austin, as we know, discovered a category of verbs of action which in certain usages do not serve to *describe* the action in question but to perform it by means of language. Saying 'I promise', under certain conditions, *is* to promise, whereas saying 'I walk' is never to walk. In this version of his theory the word 'performative' designates a class of verbs. They are performatives only in certain of their usages (for the first person in the present indicative). Since they are actions, the usages of performatives may, like all other actions, be successful or not: they can never be true or false.

The existence of implicit performatives upsets this initial conception of performative and obliges Austin to reformulate his theory.

Saying 'I state that the Earth is round' is *ipso facto* to state that the Earth is round, that is, to give to one's remark the status of a statement. Now the sentence 'I state that the Earth is round' is, *prima facie*, semantically equivalent, i.e. synonymous, with the sentence 'The Earth is round'. How could we account for that synonymy? There are two known ways in the literature of answering this question. The former is Stenius' classification in terms of moods and sentence radicals, the second is the so-called performative hypothesis. I will examine both ways, criticize them and finally offer a new solution.

4. AN EXAMINATION OF SEARLE'S NOTION OF PROPOSITION

In *Speech Acts* (1969), Searle defines proposition as that which remains invariant when the illocutionary force of a speech act varies.

Whenever two illocutionary acts contain the same reference and predication, provided that the meaning of the referring expression is the same, I shall say the same proposition is expressed. Thus, in the utterances of all of 1–5, the same proposition is expressed.[9]

He refers to the following utterances:

1. Sam smokes habitually.
2. Does Sam smoke habitually?
3. Sam, smoke habitually!
4. Would that Sam smoked habitually!
5. Mr. Samuel Martin is a regular smoker of tobacco.

And similarly in the utterance of:

6. If Sam smokes habitually, he will not live long.
7. The proposition that Sam smokes habitually is uninteresting.

the same proposition is expressed as in 1–5, though in both 6 and 7 the proposition occurs as part of another proposition. Thus *a proposition is to be sharply distinguished from an assertion or statement of it*, since in utterances of 1–7 the same proposition occurs, but only in 1 and 5 is it asserted.[10]

Searle claims that not only the same propositional content is common to 1 and 6, but also to 3. He concludes:

I might summarize this part of set of distinctions by saying that I am distinguishing between the illocutionary act and the propositional content of the illocutionary act ... .

The reader familiar with the literature will recognize this as a variation of an old distinction which has been marked by authors as diverse as Frege, Sheffer, Lewis, Reichenbach and Hare, to mention only a few.[11]

Thus, Searle treats the following three contrasts as variants of the *same* distinction:

(a) Frege's well known division of an assertion into a sentence and a sign for assertion (⊢). As an example, consider:

If the light is green, then the cars may go.

Here, 'the light is green' and 'cars may go' are unasserted propositions while the whole conditional sentence is asserted.

(b) The contrast introduced by Hare, between neustics and phrastics, and illustrated by the following example from Hare:

Sugar in my coffee   please
Sugar in my coffee   yes
phrastics   *neustics*

(c) The contrast introduced by Austin between locution and illocutionary force illustrated by the following example:

    You are going to leave.   statement
    You are going to leave!   command
    You are going to leave?   question
    locution   *illocutionary force*

Searle presents as superficial, differences which will prove to be important. As Searle's conflation of these three distinctions conceals serious problems which spring out of his theory of proposition, it is worth dwelling on this matter and trying to clear up his confusion.

There is something important in common between an asserted and an unasserted proposition, in the sense in which Frege uses these terms: for example, between 'the light is green' occurring as antecedent of a hypothetical proposition, and 'the light is green' occurring as categorical proposition in a *modus ponens* inference. We have two examples of the *same* sentence, expressing the *same* proposition first unasserted, then asserted. Contrary to this, an imperative sentence and a declarative sentence made up of the *same words* and, analysed following Hare's method, have only the *phrastic* element in common. Now, this phrastic element is *less* than a sentence. It is the sentence devoid of its copula. If the phrastic is less than an unasserted sentence, the neustic is *more* than the assertion sign, which shows that Hare's distinction does not coincide with Frege's. In this point, Hare's subsequent developments of his original analysis confirm our reasoning. In an article entitled 'Meaning and Speech Acts' (1970),[12] which appeared after Searle's work, Hare recognizes that what he initially called neustics cover two things: the *tropic* or grammatical mood and the *neustic* in the strict sense which corresponds rather well to Frege's assertion sign.

Let us now compare contrast (a) with contrast (c).

Frege's distinction between asserted and unasserted propositions differs also from Austin's distinction between locutionary act and illocutionary force. Consider the sentence 'you leave' uttered in a declarative, imperative and interrogative tone of voice, respectively; that means with the illocutionary force of a statement, a command and a question. What do these three speech acts have in common? A sentence? Yes, if one takes 'sentence' as a syntactic entity. Searle means something more, however, since he uses the word 'proposition'. But it is hard to pin-point this common element. One might even argue that it is just as hopeless to capture a common propositional content in the three sentences under consideration as it is to capture one in the two sentences

> The bank (of the river) is far from here.
> The bank (financial institution) is far from here.

I will answer this objection and offer a solution to that question. Before spelling this out, I would like to review two other attempts at answering the same question: Stenius' analysis and the Performative Hypothesis.

5. STENIUS' ANALYSIS

In *Mood and Language-game* (1961), Stenius follows a suggestion made in *Philosophical Investigations* and analyses sentences into a mood and a sentence radical. For instance the three sentences

> You live here now
> Live (you) here now!
> Do you live here now?

decompose into

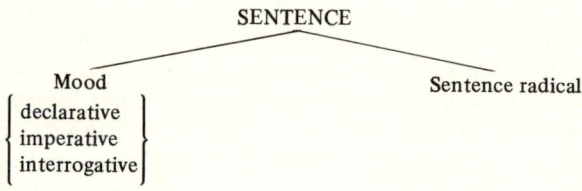

Stenius explains the distinctive role of these two components with an analogy borrowed from Wittgenstein:

To take Wittgenstein's example the showing of the picture of a boxer may have the function of indicating to somebody how a certain man stood in a certain situation – this is an indicative function – or the function of showing how a person shall (ought to) stand in a certain situation – this is an imperative function – or, finally, the function of presenting a state of affairs concerning which the person to whom the picture is shown has to decide whether it obtains or not – this is an interrogative function. *When one utters* a sentence, the sentence radical corresponds to the picture, whereas the modal element indicates the function.[13]

In this view, there is a common content to statements, commands and questions and this is the *sentence radical*.

The method of sentence radical has been criticized by David Lewis in 'General Semantics' (1972), in these terms: "It (the method of sentence radicals) works well for declaratives, imperatives, and yes-no questions. It is hard to see how it could be applied to other sorts of questions, or to sentences like 'Hurrah for Porky'". What I find objectionable in Stenius' theory is something different: the key notion of sentence radical has not been sufficiently explained and justified.

6. THE PERFORMATIVE HYPOTHESIS

According to Ross in 'On declarative sentences' (1970), a sentence like

>Prices slumped

has the structure given below

>((I) ((V) (you) (((prices) (slumped)))))
>S NP VP V NP NP S NP VP
>
>>+ V
>>+ Performative
>>+ communication
>>+ linguistic
>>+ declarative

In other words, the surface structure of 'prices slumped' corresponds to the deep structure

>I declare that prices slumped.

We can thus very neatly isolate the common element shared by the following three sentences:

>You are going to leave. (statement)
>You are going to leave! (command)
>You are going to leave? (question)

The common element is expressed by the subordinate clause in the three following performative sentences which are supposed to be *semantically equivalent* to the three former sentences

>I state that you are going to leave.
>I command that you are going to leave.
>I ask whether you are going to leave.

In French this would apparently not work so well, since the verb of the subordinate clause changes in accordance with the mood:

> J'énonce que vous *partirez* incessamment.
> J'ordonne que vous *partiez* incessamment.

This objection, however, is not fatal. The proponents of the performative analysis would reply that the difference in mood in French is simply a redundant expression of what is already lexically expressed by the performative verb.

There are, however, other objections which cannot be so easily dispelled. In 'On Performative Sentences' (1976), Gazdar enumerates eight consequences of the performative hypothesis[16] which can be proved to be false. Let us consider, for instance, the two following sentences

> I declare that the earth is flat.
> The earth is flat.

The proponents of the Performative Hypothesis would contend that they are *semantically equivalent*, that is to say *synonymous*. As an argument, they would point to the fact that if the hearer were to reply 'That is true', he would mean 'It is true that the earth *is* flat' in both cases and not 'It is true that you *declare* that the earth is flat' in the first case. This argument, however, is not compelling. As Gazdar observes, one might argue that 'That is true' used as a reply to the first sentence is ambiguous between 'It is true that the earth is flat' and 'It is true that you *declare* that the earth is flat and true that the earth is flat'.

7. HAUSSER'S TREATMENT OF MOODS

In 'Surface Compositionality and the Semantics of Mood' (1978) Hausser tackled the same problem.[17] If we treat as *semantically equivalent* these two sentences:

> I order you to leave
> Leave!

he notices that we will be committed either to Lewis' view, according to which 'Leave!' is a proposition because, 'I order you to leave' is one — a very counter-intuitive thesis indeed, or to Austin's view, according to which neither one nor the other of these sentences expresses a proposition since the second obviously does not. The latter view has the unwanted consequence of 'breaking the paradigm' by offering a different account of

> I order you to leave (not a proposition).
> You order me to leave (proposition).

Hausser rejects the claim that sentences like 'I order you to leave' and 'Leave!' are *semantically equivalent*. He concedes only that they may be used in locutionary acts which are equivalent on the *illocutionary level* i.e. that they are *pragmatically equivalent*. The semantics he provides clearly display the semantic differences between the two sentences. To the former, he ascribes as a denotation a proposition in the Montague sense, i.e. a function from possible worlds to truth-values. To the latter he ascribes a property, namely the property which the speaker wants the hearer to acquire, i.e. a function from possible worlds into functions from the domain into truth-values.

Hausser is certainly right when he says that the verb 'order' has the *same semantic content* in 'I order you' and 'You order me' and that 'I order you to leave' and 'Leave!' have a *different semantic content*. He is also right in saying that 'I order you to leave' and 'Leave!' are *pragmatically equivalent*. There is nevertheless something which he misses: namely the *common semantic part* of 'I order you to leave' and 'Leave!' which Searle calls 'Proposition'. Is it possible to account for the latter without relapsing into either Stenius's analysis or into the Performative Hypothesis? I will try to take up this challenge and to give, and justify, an affirmative answer. Before doing so, I will consider an objection which seems to threaten the method I am going to use for isolating the common propositional content.

8. A VINDICATION OF SEARLE'S POSITION

In *Word and Object* (1960), Quine warns us against the fallacy of subtraction which is exemplified in the following statement "if we can speak of a sentence as meaningful... then there must be a meaning that it has".[18] Another example he gives of this fallacy reads as follows: "we could as well (going by the same argument) justify the hypostasis of sakes and unicorns on the basis of the idioms 'for the sake of' and 'is hunting a unicorn' ".[19] Quine's warning should not, however, deter us from using, as we shall, a subtraction argument. Subtraction should not be held responsible for the counter-intuitive hypostasis of 'sakes' but rather a *careless use* of it. Once one acknowledges the fact that "for the sake of" is an indivisible whole and refrain from applying a principle of compositionality which, as most linguists admit, does not work for idioms, one avoids the unwanted consequences pointed to by Quine. As for the second case, it is a *reductio ad absurdum* only for those who subscribe to Quine's syncategorematic treatment of opaque constructions, according to which 'unicorn hunting' should be taken as an indivisible whole an account which, again, runs counter to that which linguists would give and which is motivated only by metaphysical (but respectable) considerations.

Having, I hope, disposed of the objections levelled against that particular sort of argument by analogy which consists of extending to non-numerical entities the arithmetical principle of subtraction ('when you subtract equals from equals you get equals'), let us see how one might vindicate Searle's thesis that there is a proposition common to 'You are going to leave', 'You are going to leave?' and 'You are going to leave!'.

Let us consider first 'I order you to leave' and 'Leave!'. It has been agreed that these two sentences are *pragmatically equivalent*. Both are ways of performing the *same act* of commanding that you leave. A fortiori they are both examplifications of the same *kind* of act: i.e. they are both acts of commanding. In other words, they have the same illocutionary force (the force of a command as opposed to the force of a question or to the force of an assertion). Let us *subtract* the *same* illocutionary force from the *same* illocutionary act, and we are bound to obtain the *same remainder*: this is precisely what Searle calls 'proposition'. But instead of deriving it by *abstraction* from acts of *different kind* (command, question and assertion) I obtained it by *subtraction* from two ways of performing the *same act*.

When one subtracts the force from the act, one is left with the proposition. In the case under examination, i.e. in the case of the act which can be performed either by saying 'I order that you leave' or 'Leave!', the *act* is the *same* (the act of ordering you to leave), the *force* is the *same* (it is a command) and the *proposition* obtained by subtraction is the *same* (the proposition that you leave). Yet it is possible for the two sentences to be *semantically different*. This surprising fact can be understood once it is recognized that the *pragmatic criterion* of propositional identity is less sharp than the *semantic criterion*.

From a pragmatic point of view, 'I order you' does not belong to the propositional content. From a semantic point of view it does. Once this is taken into account one understands the subtle relationship which obtains between 'I order you to leave' and 'Leave!'. It is true that these two sentences *share* a proposition, expressed by 'that you leave', but the point is that this proposition *does not exhaust* the semantic content of 'I order you to leave'. The latter sentence involves not *one* proposition but *two*. The second *proposition* the one expressed by 'I order you', however, is likely to escape the speech act theorist's attention since, the self-verifying sentence 'I order you to' is *pragmatically used*, not as *proposition bearer* but as an *illocutionary force indicating device*. *Semantically*, however, it is a proposition bearer.

9. A NEW ACCOUNT OF SEARLE'S CONCEPT OF PROPOSITIONAL CONTENT

The relationship which obtains between 'I order you to leave' and 'Leave!' can, to a certain extent, be compared with the relationship which obtains between ' "Snow is white" is true' and 'Snow is white'. In both cases, the former sentence contains two propositions, one of which is referred to or mentioned and the other of which is used, whereas the second sentence of each contains one proposition only, which is used. On the other hand, in the two pairs, one member of the pair stands in some sort of equivalence with the other: pragmatic equivalence in one case, material equivalence in the other. Yet, in neither case does this equivalence amount to synonymy.

Earlier in this chapter we were considering the nature of Searle's propositional content and raised objections against identifying it either with Frege's unasserted proposition, with Hare's phrastics, or with Austin's locutionary act. As I have indicated, I propose a fourth candidate: Searle's propositional content should be conceived as the meaning of a clause of the form 'that p', i.e. of a sentence *described*, as opposed to a sentence *used*.

On the basis of my analysis, one can do justice to Hausser's claim that the two sentences 'I order you to leave' and 'Leave!' are not synonymous and at the same time do justice to Searle's claim that they share the same propositional content. One problem remains to be solved, however, before we can claim to have reconciled the two theories: with regard to the point of view that I am presenting, a *proposition* (the semantic correlate of 'that you leave') can be *part* of the meaning of an imperative sentence, but, if we follow Hausser, the meaning of an imperative sentence is not a *proposition* but a *property*. How is it possible for a proposition to be part of a property? The way out of that puzzle is not hard to find. Let us simply bear in mind that Searle's and Hausser's approaches do not belong to the same level. Hausser offers an analysis of the *surface form* of the two sentences under consideration; Searle offers an analysis of their *deep structure*.

The importance of producing an interpretation which enables us to combine Searle's insights with Hausser's insights lies in the fact that their respective theories will prove to be very important in other areas: Hausser's for a satisfactory treatment of semantics (see Chapter VII) and Searle's in connection with the problem of defining a criterion of propositional identity (see Chapter IX).

From the point of view of nominalism, the entity introduced by Searle's notion of proposition is a *nominalized sentence* in the deep structure. Admittedly, the proposition is not that linguistic entity, but its meaning.

CHAPTER IV

Searle's notion of proposition, however, does not prejudge any specific account (nominalist or realist) of this meaning. Up to now we are committed to a new abstract entity, i.e. a sentence in deep structure; but the problem of the status of its meaning remains open.

REFERENCES

[1] W. E. Johnson, *Logic*, 1921, Dover, New York, 1964, Part. I, p. 3–4.
[2] W. E. Johnson, *Ibid.*, p. 4.
[3] W. E. Johnson, *Ibid.*, p. 3.
[4] J. Lyons, *Introduction to Theoretical Linquistics*, Cambridge University Press. 1968, p. 439.
[5] P. T. Geach, 'Assertion', *Philosophical Review* 74, (1965) 449.
[6] Geach, *Ibid.*, p. 450.
[7] G. Ryle, 'If, So, and Because', in M. Black (ed.), *Philosophical Analysis*, Cornell, Ithaca, 1950, pp. 323–340.
[8] J. Searle, *Speech Acts*, Cambridge University Press, 1969, p. 3.
[9] Searle, *Ibid.*, p. 29.
[10] Searle, *Ibid.*, p. 29.
[11] Searle, *Ibid.*, p. 30.
[12] R. M. Hare, 'Meaning and Speech Acts, *Philosophical Review* 79, (1970) 20–21.
[13] E. Stenius, 'Mood and Language-Game', *Synthese* 17, (1967), reprinted in J. W. Davis, D. J. Hockney and W. K. Wilson (eds.), *Philosophical Logic*, Reidel, Dordrecht, p. 253.
[14] D. Lewis, 'General Semantics', *Synthese* 20, (1970), reprinted in D. Davidson and G. Harman (eds.), *Semantics of Natural Language*, Reidel, Dordrecht, p. 207.
[15] J. Ross, 'On Declarative Sentences' in R. A. Jacobs and P. S. Rosenbaum (eds.), *Readings in English Transformational Grammar*, Waltham, Ginn/Blaisdell, 1970, pp. 254–261.
[16] G. Gazdar, 'On Performative Sentences', *Semantikos* 1 (1976) 37–62.
[17] R. Hausser, 'Surface Compositionality and the Semantics of Mood', *Amsterdam Papers in Formal Grammar*, Vol. II, 1978, p. 176 (pp. 174–193).
[18] W. V. O. Quine, *Word and Object*, Holt, New York, 1960, p. 206.
[19] Quine, *Ibid.*, p. 207.

CHAPTER V

THE NATURE OF FACTS

1. THE NATURE AND STATUS OF FACTS IN RUSSELL'S 'PHILOSOPHY OF LOGICAL ATOMISM'

In 'The Philosophy of Logical Atomism' (1918), Russell frees himself from the Platonism to which he had subscribed earlier. He no longer regards propositions as entities which exist independently of thoughts and sentences. He now identifies them with the symbols that express them:

"A proposition is just a symbol".[1]

"A proposition ... is a sentence in the indicative ...".[2]

"If you were making an inventory of the world, propositions would not come in".[3]

"... obviously propositions are nothing".[4]

Does this mean that Russell has become a nominalist so far as propositions are concerned? One cannot assume this. In fact, if Russell eliminates propositions from the extra-linguistic world, he introduces, in return, new entities in their places: *facts*. These have so much in common with propositions that one is tempted to see them as vestigial propositions.

The following passage fosters this impression:

What I call a fact is the sort of thing that is expressed by a whole sentence, not by a single name like 'Socrates'.[5]

'The first thing I want to emphasize is that the outer world ... is not completely described by a lot of 'particulars', but that you must also take account of these things that I call facts, which are the sort of things that you express by a sentence, and that these, just as much as particular chairs and tables, are part of the real world.[6]

Thus, it may not surprise us to see Quine, in 'Russell's Ontological Development' (1966), contest Russell's claim that he has freed himself of propositions:

As for propositions, in particular, we saw Russell in this essay taking them as expressions part of the time and part of the time simply repudiating them. Dropping then the ambiguous epithet, we might take this to be Russell's net thought: there are no nonlinguistic things that are somehow akin to sentences and asserted by them.

But this is not Russell's thought. In the same essay he insists that the world does contain nonlinguistic things that are akin to sentences and asserted by them; he merely does not call them propositions. He calls them facts. It turns out that the existence of nonlinguistic analogues of sentences offends Russell only where the sentences are false.[7]

The question whether the notion of fact is more acceptable than that of proposition is not only of historical but also of topical interest.

Russell, indeed, persistently uses the word 'fact' in the definition of truth. Speaking of belief, he writes in *Human Knowledge, Its Scopes and Limits* (1948): "... in the case of true belief there is a fact to which it has a certain relation, but in the case of a false belief there is no such fact"[8]

In *My Philosophical Development* (1959) he finally confirms this usage of the word 'fact':

But whether there is one verifier or there are many, it is always a fact, or many facts, that make the statement true or false as the case may be; and the fact or facts concerned, except in a linguistic statement, are independent of language and may be independent of all human experience.[9]

It is not certain however that the word 'fact' in the two works just mentioned has the same sense as in 'The Philosophy of Logical Atomism'. We cannot say for sure from the context that Russell in his later writings intends that the word 'fact' have the same sense that it has in 'The Philosophy of Logical Atomism'; that is, the sense of "a state of affairs formed by the union of individuals and relations", or whether, on the contrary, he gives it the usual and vague sense that connotes the notions of reality and event.

In the case of either hypothesis the use of this word is open to criticism. In the technical sense, 'fact' is opposed to 'event' and 'object'. Used in this way, the word raises serious objections that we shall examine. In its usual sense, its lack of precision makes it incapable, in turn, of expressing a philosophical adherence to a realism of a well-determined kind like logical atomism; it simply expresses an adherence to a rather vague epistemological realism.

2. THE MERITS OF RUSSELL'S NOTION OF FACT

Before criticizing the concept of 'fact' that Russell introduces in 'The Philosophy of Logical Atomism' (1918), fairness requires that we discuss its merits, especially since that work contains elements of truth that any theory of the proposition must be able to incorporate.

As we saw, in 1918 Russell rejects propositional entities, but admits the existence of facts. This gave Quine the idea that Russell had changed his language rather than his ontology. One must, however, recognize that he reformed his ontology *as well*, at least *quantitatively*. He accomplishes this in freeing himself of *false* propositions.

If the difference between propositional entities and facts were only *quantitative*, one might perhaps contest its metaphysical import. But it is more than this; it reflects a *structural difference*.

Speaking of propositions reduced to symbols, Russell writes: "... there are *two* propositions corresponding to each fact".[10] This comes close to saying that every fact verifies only *one* of the *two* sentences of opposed sign (affirmative or negative) that one may construct from the elements of an atomic proposition (names, predicates with one or more places).

From this it follows that the *semantic* relation uniting a *sentence* with a fact that verifies it or makes it false cannot be confused with the semantic relation linking a *name* to a given object. This dissociation, as we shall see, is an important advance which did not receive its just due.

In an ideal language, that is an exhaustive one and one without synonyms and homonyms, there would be a *one-to-one* relationship between individuals and their names, that is, to an open class of words of the same kind. The semantic relation that unites sentences to facts, on the contrary, would have a different logical form. It would be a *many-one* (two-to-one) relationship.

This is, we believe, a logical difference of great importance. It allows us, in effect, to bring into full light that which distinguishes sentences from names and to recognize the character of the former. It took some time for logicians to become aware of this distinction. They did not sufficiently scrutinize the valuable Aristotelian contrast between the proposition, which may be true or false, and the concept, which can be neither true nor false.

For Hobbes, for example, the sentence was a complex name. In 'Ueber Sinn und Bedeutung' (1895), Frege says that the sentence is a name of its truth value. Church does the same in *Introduction to Mathematical Logic* (1956). Carnap, finally, adds nothing new to this point in *Meaning and Necessity* (1947, 1956). Indeed, he classifies sentences as designators which, like names, have an extension and an intension. Certainly, he recognizes a different extension for sentences from that which he attributes to names and predicates (instead of being an individual or a class, this extension is a truth value), but this distinction does not seem to be sufficient. In fact, it masks an important asymmetry, the asymmetry between true and false. The false is parasitic in relation to the true, whereas the null class is not parasitic in relation to other classes.

The *sui generis* character of propositions, thus, is undisputably obscured in Frege's, Carnap's and Church's theories. In ignoring the asymmetry to which we alluded, these theories make inexplicable the priority, the 'normativity', of the true relative to the false. Russell's theory, on the contrary, provides us with the first elements of the conceptual apparatus necessary for the description of this asymmetry: there is only *one* fact for each *pair* of contradictory propositions. One of these propositions, therefore, must *give way* to the other which must take precedence.

It remains to explain why one such proposition rather than the other does have this priority. Even if Russell's theory of 1918 provides no (or an incomplete) answer to this problem, at least it has the merit of having caused us to *pose* it, something the other semantic theories inspired by Frege do not do. Wittgenstein's picture theory in the *Tractatus* (1922), may be considered as an attempt to *resolve* the present problem. We shall have occasion to return to this point in a later chapter.

For all these reasons we believe that Russell's theory of facts — which locates the difference between sentences and names or predicates in the *nature* of the semantic *relation* rather than in the *category* to which the second *term* of the relation belongs — is much healthier in many respects than a number of semantic theories currently in vogue among logicians. Perhaps these latter theories are adequate to resolve particular problems in formal semantics; but if one wishes to construct a general theory of the proposition in which its *function* is explained, then one will have to account for the *sui generis* character of the proposition. Russell's theory of logical atomism, which does do this, consequently contains elements indispensable to the elaboration of a satisfactory theory of propositions. But it also contains defects which we shall now point out.

3. THE DEFECTS OF RUSSELL'S THEORY OF FACTS

One may, however, disagree with Russell's theory of facts on two scores: one may object to his reification of the *distinction* between atomic and non-atomic facts and to his reification of *facts* themselves.

Our first thesis does not require extensive justification. Today, everyone grants that the gap between atomic and non-atomic facts is not inscribed in the texture of reality, but that it is relative to the subtlety of our recognition patterns and to our *language*.

Ayer has said:

... it is a mistake to think of the facts as being 'out there' waiting to be known. For what is to be regarded as a fact will partly depend on what symbols we use. This is not to say that it will not also depend upon certain non-symbolic occurrences. ... My point is only that although what occurs may be logically independent of the use of symbols, what is known is not.[11]

One may add to Ayer's declaration the following passage by Louis Boisse that Bachelard would not have disputed:

The fact is much less something which has been verified than a construction of the mind. Rigorously speaking, facts do not exist already made as clothes in a clothes shop and the role of the scholar is not limited to calling them up in turn following the needs of his discipline, but rather creating them in some way by isolating them from the whole complex of which they form part. – It must be said that this creation is neither artificial nor arbitrary.[12]

4. WITTGENSTEIN'S CONCEPTION OF FACT

In his philosophy of logical atomism Russell reifies facts. We maintain that this deviation from the principle of methodological nominalism is not justified. This is our second thesis. In order to establish it, we shall present an alternative solution to the ontological conception of facts, a conception which one may call transcendental or metalinguistic. Then we shall try to refute the philosophical arguments invoked to establish the existence of facts. Finally, we shall present arguments from transformational grammar in support of our thesis.

We find the solution, which replaces the ontological conception of facts, in Wittgenstein's *Tractatus*. Our position in this matter is, therefore, opposed to Quine's, who finds a similarity between Russell's ontology in 'The Philosophy of Logical Atomism' and that of the earlier Wittgenstein.

Russell's ontology of facts here is a reminder of Wittgenstein, but a regrettable one. Wittgenstein thought in his *Tractatus* days that true sentences mirrored nature, and this notion led him to posit things in nature for true sentences to mirror; namely, facts.[13]

The connection that Quine sees between Russell's and Wittgenstein's concept of facts does not appear acceptable to us. In fact, it is in contradiction to certain passages of the *Tractatus*. Indeed, Wittgenstein writes at the beginning of the treatise: "the world is the totality of facts, not of things". This statement does not mean the same as Russell's parallel statement: "The world contains *facts*".

In 4.1272 Wittgenstein writes:

Thus the variable name '*x*' is the proper sign for the pseudo-concept *object*.

Wherever the word 'object' ('thing', etc.) is correctly used, it is expressed in conceptual notation by a variable name.

Wherever it is used in a different way, that is as a proper concept-word, nonsensical pseudo-propositions are the result.

So one cannot say, for example, 'There are objects', as one might say, 'There are books'. And it is just as impossible to say 'There are 100 objects', or, 'There are $\aleph_0$ objects'.

And it is nonsensical to speak of the *total number of objects*.

The same applies to the words 'complex', 'fact', 'function', 'number', etc.[14]

As Kant cautioned us against any application of the concept of *causality* to the things in themselves, Wittgenstein here warns us of the disastrous consequences attending an ontological use of the expression 'concrete fact'. This warning seems to imply that Wittgenstein's statement about the world and facts may not be interpreted as ontological theses analogous to Russell's, but as *transcendental affirmations*, with this difference, that they concern *language* and not *understanding*.

If one stresses the qualification *pseudo-concepts* that Wittgenstein applies to the concept of a *concrete fact, number*, etc..., one comes very naturally to assimilate them to *meta-concepts*, that is, to concepts of the *metalanguage*. Without doubt, Wittgenstein does not go that far. This is not because our transition from *pseudo-concept* to *meta-concept* is not a natural transition, but rather because it involves a notion that Wittgenstein would not accept: the notion of a metalanguage. The 'epistemological obstacle' which stops him here is his complete and exclusive adherence to what is conveniently called a picture-theory.

The step that Wittgenstein, held back by his theory, refused to take was, as we know, briskly taken by Carnap. In his *The Logical Syntax of Language* (1934), he interpreted "the world is the totality of facts, not things" as "science is a system of sentences, not of names".[15] In this form the proposition from the *Tractatus* no longer has an ontological significance, and the break with Russell's parallel thesis is achieved.

In Carnap's interpretation, the resemblance between Kant and Wittgenstein is accentuated to such a degree that it gave J. Vuillemin the idea that the first proposition from the *Tractatus* formulated in the formal mode of speech was a modernized version of the first antinomy of pure reason.[16] The reasoning is as follow: "If 'the world' is a concept of second order, signifying what one calls 'science' in the *formal mode of speech*, then one sees clearly that it is absurd to ask oneself if the world is finite or infinite in time or space". "Is it or is it not finite in time or space?" is indeed a question one may ask oneself in connection with scientists but not in connection with science.

Is the transcendental, rather than the ontological, interpretation of the *Tractatus Logico-Philosophicus* which we defend faithful to Wittgenstein's intentions? Black contests this.

A positivistic rephrasing of the argument of the text, eschewing any reference outside language, would be flat and unpersuasive. There is an air of the improvised and contrived about discussion of language by writers like Carnap who have been strongly influenced by the *Tractatus*, but have altogether rejected its ontology.[17]

The linguistic interpretation of the *Tractatus Logico-Philosophicus* makes it too bloodless for Black's taste.

However, we do not have to expresses an opinion here on this difficult exegetical question. It suffices to recall that a comparison between Russell's logical atomism and that of Wittgenstein's reveals that the notion of *fact* may be understood in two ways, in an *ontological* sense and in a *metalinguistic* sense.

Therefore, the sense of our second objection to Russell may be conveniently put in precise terms. We fault Russell for having taken the concept of fact as a concept of the object-language, whereas in reality we have to do with a metalinguistic notion. Now, if one 'rectifies' Russell's thought on this point, one eliminates, at the same time, the last vestige of propositional realism burdening it. The payoff from this linguistic retreat is, therefore, considerable.

The crucial question is whether this ascent is legitimate, whether Wittgenstein (in Carnap's interpretation) is right against Russell. In order to answer this question we must appraise the force of the argument developed by the supporters of the one or the other doctrine.

5. ARGUMENTS FOR AND AGAINST THE ONTOLOGICAL INTERPRETATION OF FACTS

(A) Stebbing writes in *A Modern Introduction to Logic* (1930):

The fact *that Charles I was unfortunate* consists of two constituents – a thing (*Charles I*) and a quality (*being unfortunate*). *That Charles I married Henrietta Maria* is a fact consisting of three constituents – two things (*Charles I* and *Henrietta Maria*) and a relation that relates them (*marriage*).[18]

In 'Substance, Events and Facts' (1932), she expresses herself in terms which leave no doubt about the sense she gives to the word 'fact'. It is uncontestably an *ontological* sense. She refers to Wittgenstein, but a Wittgenstein interpreted in a realist sense.

Anxious to guarantee for *facts* an *ontological reality* distinct from that of *events* Stebbing takes after Whitehead: "I think ... that Professor Whitehead is misusing language and thus misleading us, or else he is talking nonsense, when he says that 'the facts of life are the events of life' ".[19]

In order to prove the existence of facts she advances two arguments:

(a) The atemporality of facts: "A fact is not an event. It does not occur *at* a time although some facts are facts with regard to a particular time".[20]

(b) "A complete description of the world ... would be a description in terms of facts".[21]

These two arguments cannot possibly provide a basis for an ontological conception of facts. They both prove that facts *differ* from events, which we do not dispute at all, since we attach the word 'fact' to the metalanguage and the word 'event' to the object-language. Rather, Stebbing should have tried to show that they *differ though both are on the same level*.

(B) Ramsey distinguishes facts from events. The arguments he uses to effect this notional dissociation are instructive for our purpose. In the following passage from 'Facts and Propositions' (1927), he introduces the distinction between fact and event:

> The connection between the event which was the death of Caesar and the fact that Caesar died is, in my opinion, this: 'That Caesar died' is really an existential proposition, asserting the existence of an event of a certain sort, thus resembling 'Italy has a king', which asserts the existence of a man of a certain sort. The event which is of that sort is called the death of Caesar and must no more be confused with the fact that Caesar died, than the King of Italy should be confused with the fact that Italy has a king.[22]

The distinction between fact and event is instituted with the help of an unassailable logical argument. Ramsey founds the distinction on the following contrast between the 'logical grammar' of events and that of facts.

(a) An event *may* be expressed by a *definite description*. Just as two different definite descriptions may describe the same individual, two *non-synonymous* descriptions of an event may be descriptions of the *same* event. For example, 'Caesar's death', though not synonymous with 'Caesar's assassination', describes the same event.

(b) On the contrary, a fact *must* be expressed by a *proposition*. Now, in contrast to what applied to descriptions, two coextensive but *non*-synonymous sentences may not express the *same* fact; so close is the affinity between a fact and a sentence expressing it. For instance, 'that Caesar is dead' and 'that Caesar has been assassinated' cannot express the same fact, because the two sentences that purport to state it are not synonymous.

To sum up, what constituents do we find in facts and not in events? Could it be an ontological ingredient? Not at all. What distinguishes facts from events is solely a logical ingredient. Furthermore, Ramsey says that a fact is "an event and also a character". In other words, a fact is an event *apprehended through a description*, a description that is involved in the definition of the fact, whereas it is not involved in the definition of an event.

The way in which Ramsey demonstrates the basis of this notional distinction is very convincing, but it does not at all contradict our rejection of facts. The fact whose existence Ramsey succeeds in establishing is far from having an existence *independent* of language. It is, on the contrary, inseparable from language, and furthermore does not constitute an intolerable realist residue like the 'fact' of Russell's conception.

(C) Ducasse and Strawson, who reject an ontological interpretation of facts, defined them in terms of true statements. It has been argued against this version of the linguistic theory of facts that it cannot account for the *causal* force attached to *facts* and which is absent from *statements*. One may indeed say that a fact *caused* another fact, but not that a true *statement causes* another statement.

In 'Facts, Events and True Statements' (1966), F. A. Tillman advances the following counter-example to the equation 'Fact = true statement':

Consider... 'The Fact that the Japanese Navy was defeated made it possible for the United States to invade Mindanao'. If we substitute 'The true statement that' for 'The fact that' we obtain the sentence 'The true statement that the Japanese Navy was defeated made it possible for the United States to invade Mindanao', which is not a paraphrase of the original.[23]

Tillman's objection had already been formulated by M. T. Keeton by way of another example in 'On Defining the Term 'Fact'' (1942)[24].

We readily concede to these authors that the word 'fact' *is not always* a concept of the metalanguage synonymous with 'true statement'. But this does not imply, as one might think, that we are *ipso facto* forced to admit that this word is *sometimes* a concept of the object-language designating an ontological entity *sui generis*, a physical entity with a structure similar to that of propositions.

Certainly, the word 'fact' is *equivocal*, as E. L. Beardsley notes in 'The Semantical Aspect of Sentences' (1943),[25] but it never designates an entity distinct from objects and having a quasi-grammatical structure. When it is not synonymous with 'true statement' it designates an event.

In order to establish this thesis, it is necessary that we prove that *ambiguity* is not attributed to the word 'fact' for the sake of argument. This proof is possible today, thanks to the resources of transformational grammar, which detects ambiguities by *transforming* suspect expressions and inserting them into syntactic contexts that play the role of 'detectors' much like litmus paper in chemistry.

6. APPLICATION OF METHODS OF GENERATIVE GRAMMAR TO DETECT THE ONTOLOGICAL NATURE OF FACTS

In 'Unfair to Facts' (1954), Austin, in opposition to Strawson, tries to prove that facts exist in the world with the same status as events. His tactic consists in looking for cases where facts coincide with events, which unites them *ontologically*. His tactic is illustrated in the following two examples given by Tillman in 'Fact, Events and True Statements' (1966).

(A) "Phenomena, events, situations, states of affairs are commonly supposed to be genuinely-in-the-world, and even Strawson admits events are so. Yet surely of all of these we can say that they *are facts*. The collapse of the Germans is an event and is a fact — was an event and was a fact."[26]

(B) "But surely we can witness facts? And observe them and have personal acquaintance with them."[27]

In order to resist Austin's argument in favor of a *realist* conception of facts, the obvious parry would be to dissociate what he united by demonstrating that he could unify event and fact only by virtue of an *ambiguity* in the contentious word. As we shall see, transformational grammar provides us with the tools for such a demonstration.

In *Linguistics in Philosophy* (1967), Z. Vendler exploits the resources of modern linguistics to refute Austin. He shows that the same expression may at times express a fact, at other times an event, depending upon context, but it may not simultaneously express both. Consider the following two sentences:

    John's death was painful.
    John's death was surprising.

The *transformations* of these two sentences are fundamentally different. The same is true for:

The collapse of the Germans was gradual
The collapse of the Germans was denied.

One may say 'The collapse of the Germans took place in 1944', but not 'That the Germans fell took place in 1944'. Likewise, one may say 'He believes that the Germans collapsed', but not 'He believes in the collapse of the Germans'.

The ambiguity in the circumlocution 'The collapse of the Germans' defeats Austin's reasoning, which rests upon a comparison of sentences chosen on the basis of their surface structure. Vendler writes:

> As it by no means follows that since the collapse was a gradual or bloody event, the fact of that collapse has to be gradual or bloody, and as it by no means follows that since the fact of that collapse has been denied or contradicted, any event has to be denied or contradicted, so it is equally absurd to conclude that since the collapse of the Germans was an event that took place in the world, any fact has to take place or simply be in the world. Austin's syllogism has four terms.[28]

7. WHY THERE CANNOT BE FACTS

In *Meaning and Necessity* (1947, 1956), Carnap endeavors to explain why a sentence may be *false* and yet have an *objective sense*, a sense that is independent from mental images, subjective states and beliefs. He proposes the following solution:

> Any proposition must be regarded as a complex entity, consisting of component entities, which, in their turn, may be simple or again complex. Even if we assume that the ultimate components of a proposition must be exemplified, the whole complex, the proposition itself, need not be.[29]

An example will make this clearer: Suppose that in the universe there are black objects which are not tables and also tables which are not black but that there are no black tables. Under this hypothesis the constituents of 'The table is black' would be exemplified, but not the complex in its totality.

Which is the faulty complex? Is it the *fact* that the table is black? Carnap does not tell us. In 'Propositions and Sentences' (1950), P. Marhenke attributes this interpretation to Carnap and criticizes it severely:

> ...the fact that exemplifies the proposition that the table is black is the black table ... in addition to the black table there is not also another object which is *the fact* of the table's being black.[30]

One may notice here a convergence — which is very significant for the theory we are trying to construct — between the results Marhenke is leading up to in his analysis of the concept of facts and those arrived at by Davidson in his analysis of the notion of truth in 'True to the Facts' (1969). In both cases the *reference* of sentences does not consist of facts, but of individuals or ordered n-tuples of individuals.

Marhenke's objection must be compared to *certain* declarations of Aristotle, who saw the problem and cautions us against the reification of facts (as it came to be called). In the *Metaphysics*, he distinguishes carefully being as truth from being as substance. He writes:

> As for 'being' *qua* truth, and 'not being' *qua* falsity, since they depend upon combination and separation.... By 'combining or separating in thought' I mean thinking them not as a succession but as a unity for 'falsity' and 'truth' are not in *things* ... but in *thought*...[31]

Another passage contrasting being expressed by the copula to being expressed by the use of substantives is even clearer:

> But since the combination and separation exists in thought and not in things, and this sense of 'being' is different from the proper senses (since thought attaches or detaches essence or quality or quantity or some other category), we may dismiss the accidental and real senses' of being.[32]

It is clear that for Aristotle the formulation of propositions by the use of two *separate* terms, the subject that designates substance and the predicate that expresses accident, is a *function* of *discursive thought*, or, as one would say today, a *function of language*. In reality, accidents are not *separate entities associated* with substances to form *facts*. Contemporary philosophers using the word 'fact' in an *ontological* sense should have pondered these texts.

Did Carnap really fail to recognize Aristotle's warning and conceive of *fact* as a reality distinct from *things*? One cannot be sure. Carnap says that in a false sentence the components are exemplified but the complex itself is not, which seems to suggest that for him the difference between true and false is *structural* rather than *ontological*. If a sentence is false, no *fact* is missing from the collection of irregular objects constituting the universe; rather, falsity arises from the fact that a linguistic *combination* does not correspond to the right *combination* in reality, an adequate predicate was not joined to a suitable subject.

But when the sentence is true, is it not the case that an appropriate combination is exemplified over and above the entities it combines? Suppose we consider the extension only. Take an atomic sentence such as

Alfred is taller than Bill.

One can correlate Alfred and Bill with the individuals they denote and 'taller than' with the set of ordered pairs made out of two individuals related by the relation 'taller than'. These extensions hold whether the sentence is true or not.

Let us suppose that the sentence is true. It seems that something more has to be introduced that would have been missing had the sentence been false. In other words, we expect that there is something out there which corresponds to the membership-relation which obtains between the ordered pair ⟨Alfred, Bill⟩ and the set of pairs ordered by the relation *'taller than'*. This extra entity is what we call a *fact*.

Such a view loses much of its appeal when we look more closely at the sentence which can be understood as 'the pair ⟨A, B⟩ is a member of the set of pairs {⟨A, B⟩, ⟨C, D⟩, etc...}'

We see at once that the same pair ⟨A, B⟩ occurs both on the left side and the right side of the membership relation sign ... $\in$ {– – –}. Now this twofold occurrence of the pair ⟨A, B⟩ in this context is meant to express the identity of the pairs of objects denoted by the pairs of names of object. If, however, we subscribe to Küng's claim that "A reflexive relation [like identity] is not a real relation, but rather a relation of reason based on the conceptual duplication of an entity that in reality is a single entity"[33] we shall not be seduced into thinking that there are facts over and above individuals and set of individuals.

If this argument were criticized for ignoring relationships in intension which free statements like 'Alfred is taller than Bill' from the triviality which accompanies their extensional readering, I should simply point to the fact that intension is brought into the picture in order to deal with the question of how to account for the *informativeness* of statements. And this is an epistemological, rather than an ontological question. The enriched picture which follows — for instance the analyses of intension in terms of function from possible worlds into extension — does not affect the analysis given above in which the sentence is viewed as describing in so many words what is unified in reality.

REFERENCES

[1] B. Russell, 'The Philosophy of Logical Atomism', in R. C. Marsh (ed.), *Logic and Knowledge*, Essays 1901–1950, Allen and Unwin, London, 1956, p. 185.
[2] Russell, *Ibid.*, p. 185.
[3] Russell, *Ibid.*, p. 214.
[4] Russell, *Ibid.*, p. 223.

CHAPTER V

⁵ Russell, *Ibid.*, p. 182–183.
⁶ Russell, *Ibid., passim.*
⁷ W. V. O. Quine, 'Russell's Ontological Development', *Journal of Philosophy*, (1966) 664.
⁸ B. Russell, *Human Knowledge*, Allen and Unwin, London 1948, p. 165.
⁹ B. Russell, *My Philosophical Development*, Allen and Unwin, London 1959, p. 186.
¹⁰ B. Russell, *The Philosophy of Logical Atomism*, p. 187.
¹¹ A. J. Ayer, *Thinking and Meaning*, 1947, p. 20.
¹² Lalande, *Vocabulaire technique et historique de la philosophie*, edn. 1959, p. 334.
¹³ Quine, *Op. cit.*, p. 664.
¹⁴ L. Wittgenstein, *Tractatus Logico-philosophicus* 4. 1272, Routledge and Kegan, London, 1961, p. 11.
¹⁵ R. Carnap, *The Logical Syntax of Language*, Routledge and Kegan Paul, London, 1964, p. 303. [Originally appeared as *Logische Syntax der Sprache* (1934).]
¹⁶ J. Vuillemin, 'La référence des phrases déclaratives', unpublished.
¹⁷ M. Black, *A Companion to Wittgenstein's Tractatus*, Cambridge University Press, 1964, p. 35–36.
¹⁸ S. Stebbing, *A Modern Introduction to Logic*, London, 1930, p. 36–37.
¹⁹ S. Stebbing, 'Substance, Events and Facts', *Journal of Philosophy* 29, (1932) 311.
²⁰ S. Stebbing, *A Modern Introduction to Logic*, p. 36.
²¹ S. Stebbing, 'Substance, Events and Facts', p. 311.
²² F. P. Ramsey, 'Facts and Propositions', *Proceedings of the Aristotelian Society*, Supplementary Volume, VII, 1927, p. 156.
²³ F. A. Tillman, 'Facts, Events and True Statements', *Theoria* XXXII, 1966, p. 125.
²⁴ M. T. Keeton, 'On defining the Term 'Fact'', *Journal of Philosophy* 39, (1942) 126.
²⁵ E. L. Beardsley, 'The Semantical Aspect of Sentences', *Journal of Philosophy* (1943).
²⁶ J. L. Austin, *Philosophical Papers*, Oxford Clarendon Press, 1960, p. 104, quoted by Tillman. Oxford paperbacks, p. 156.
²⁷ Austin, *Ibid.*, p. 116, Oxford paperbacks, p. 168.
²⁸ Z. Vendler, *Linguistics in Philosophy*, Cornell University Press, New York, 1967, p. 142.
²⁹ R. Carnap, *Meaning and Necessity*, The University of Chicago Press, Chicago, 1956, p. 30.
³⁰ P. Marhenke, 'Propositions and Sentences', *Meaning and Interpretation*, University of California Press, Vol. 25, 1950, p. 289. [My italics – P.G.]
³¹ Aristotle, *Metaphysics*, 1027 620 Loeb Classical Library, transl. H. Fredernick, Harvard University Press, London, Cambridge, 1968, pp. 306–307.
³² Aristotle, *Metaphysics*, 1027 630, *Ibid.*, pp. 308–309.
³³ G. Küng, *Ontology and the Logistic Analysis of Language*, D. Reidel, Dordrecht, p. 165.

CHAPTER VI

THE PROPOSITION IN TERMS OF BELIEF

1. BELIEF AND PROPOSITION

In 'On Propositions: What They Are and How They Mean' (1919), Russell foresaw a way of approaching the problem of *defining* the proposition and of determining its ontological status that we have not yet discussed. The paper opens with the following statement: *"A proposition may be defined as: What we believe when we believe truly or fasely"*.[1] Several contemporary philosophers have preceded or followed Russell along this path full of obstacles.

It is instructive to study how these obstacles have been overcome. The first solutions proposed were insufficient and necessitated subsequent amendments which profoundly transformed the original problem and revealed connections with other problems.

The necessity of constructing a general theory meeting all of these problems is imperative. Since certain recent solutions, like Gale's in 'Propositions, Judgments, Sentences and Statements' (1967), do not satisfy all of these exigencies *together*, they prove to be completely insufficient. We shall try to outline a global solution that escapes these defects. But first we must review the work already done.

Consider Russell's definition as a working hypothesis: "A proposition is what we believe truly or falsely". This definition presents the proposition as the object of belief. It follows from this that to determine the nature of propositions it is necessary and sufficient to determine the nature of the believed object. To succeed in this one must first ask oneself what reasons there are to posit an object of belief.

The first justification invoked is the grammatical transitivity of the verb 'to believe' and other verbs like 'to desire'. Brentano, for example, the first modern theoretician of intentionality, writes in *Psychologie vom empirischen Standpunkt* (1874):

Certain feelings are unmistakably referred to objects, and language itself indicates these through the expressions it uses ... . Joy and sorrow, like affirmation and denial, love and hate, desire and aversion, distinctly ensue upon a presentation and are referred to what is presented in it.[2]

This is almost word for word the opinion Plato attributes to Socrates and his interlocutor in the *Theatetus* (189A):[3]

Socrates – And if he thinks, he thinks something, doesn't he?
Theatetus – Necessarily.
Socrates – And when he thinks something, he thinks a thing that is?
Theatetus – Clearly.
Socrates – But surely to think nothing is the same as not to think at all.

This is also Malebranche's opinion in *Recherche de la Vérité*: "To see nothing is not to see, to think nothing is not to think".[4]

Blake also is guided by language when he writes: "... to believe nothing is simply not to believe"[5], from which one may infer by contraposition, that believing is to believe something.

Moore advances a more convincing argument in *Some Main Problems of Philosophy* (1911), but it establishes a weaker thesis. He systematically varies factors involved in a belief by operating on sentence segments which figure in statements of belief of the form '*A* believes *p*'.

This procedure reveals to him that:

(1) different mental acts may have the same object. The same subject may indeed, in turn, *understand, believe* or *doubt*[6] that $2 + 2 = 4$;

(2) two acts of apprehension may be *generically* similar and differ in *what* is apprehended,[7] for example, a belief that $2 + 2 = 4$ and the belief that $4 + 4 = 8$;

(3) two persons may entertain the same belief.[8]

This systematic variation, termed by Ryle "operation with identity", allows Moore to isolate the concept of *proposition*. "What we do, I think, mean, when we say that both persons have the same belief, is that *what* is believed in both of the two different acts is the same; we mean by a belief, in fact, *not* the act of belief, but *what* is believed; and what is believed is just nothing else than what I mean by a proposition".[9]

As we said above, this argument as such establishes less than the preceding one. It allows us to *distinguish conceptually* the proposition from a mental act, but it does not allow us to establish the *independent existence* of propositional universals; at least it does not unless one takes on a supplementary premise, such as the principle of epistemological realism formulated by J. Ruytinx as follows: "Where the mind *can* distinguish elements, there *must* be elements to distinguish".[10]

Moore does not appeal to this argument. On the contrary, he improves on that of systematic variation, and this leads to unexpected results. But before we examine these we must pay some attention to the partial conclusions which were obtained up to then.

If for one or other of the reasons we have mentioned one admits that verbs such as 'to know', 'to believe', 'to desire', 'to wish' designate attitudes of mind directed toward an object, this raises the question of the ontological status of this relation as well as of this object. Is it really a relation in the proper sense of the word? If this is the case, then it cannot exist independently of its relata, its object. Here a difficulty arises. When the relation in question is the relation of 'knowing', one may assign to it relata which, in the opinion of many, will be 'facts'. Thus Ryle, in 'Are there Propositions?' (1929–1930), describes knowing as a binary relation, the first term of which is the thinking subject and the second the *known fact*:

... *in knowing* we have a sort of apprehension the expression of which takes the shape of statement, and yet the object of which is not a proposition. ...What I know is a *fact*... .[11]

But, when the relation in question is belief, desire or hope, one cannot answer in the same way and assign to these relations as relatum a fact or a real object. Indeed, the belief may be false, the desire frustrated and the hope disappointed.

2. THE PROBLEM OF FALSE BELIEFS

Two solutions are possible to solve this difficulty. We may either assign a degenerate object to the relation in the case of a false belief, or no longer consider belief a relation, which would free us of the obligation of finding an object for it. These two solutions have been proposed, and we shall evaluate their respective merits.

The first was adopted by Brentano in the above-mentioned work: the intentional object only has a psychological existence.

Every mental phenomenon is characterized by what the scholastics of the Middle Ages called the intentional (and also mental) inexistence (*Inexistenz*) of an object (*Gegenstand*), and what we could call, although in not entirely unambiguous terms, the reference to a content, a direction upon an object (by which we are not to understand a reality in this case), or an immanent objectivity. Each one includes something as object within itself, although not always in the same way. In presentation something is presented, in judgment something is affirmed or denied, in love [something is] loved, in hate [something] hated, in desire [something] desired, etc.[12]

In our opinion, the major defect in an explanation of this kind is that it too radically separates true belief from false belief. The theory of intentionality *provides an account* of the *differences* separating these two beliefs, but it *obfuscates* their *similarities*. A satisfactory theory must bear both in mind.

Moore at first adopted Brentano's way and admitted for a time the existence of intentional objects as the relata of the relation of belief. However, he recognized very soon the difficulty we have pointed out, and he changed his tune. To verify that the second solution is superior to the first we can follow out the thread of Moore's reasoning to see why he turned to this solution. His reasoning is an important landmark, as it represents definite progress toward a satisfying solution, even though he does not completely succeed in reaching his goal.

Recall that Moore allowed the elements in a statement of belief to vary; he now goes on to investigate the significance one should attribute to this latitude. Taking as examples the belief that *lions exist* and the belief that *bears exist*, he examines the differences between these two beliefs. He answers the question is these terms:

It seems difficult to see how it can consist in anything else than that the one belief has a specific kind of relation to one object, while the other has the same kind of relation to a different object.[13]

But Moore recognizes that if one construes belief as a relation in the case of true belief, one has the same reason to do so in the case of false belief. He asks us to consider the case of two children, one of whom believes that *griffins exist*, the other that *centaurs exists*.

And there is the same reason here, as in the case of true beliefs, for saying that the difference consists in the fact that the one belief has a specific relation to *the proposition* that griffins exist and the other the same relation to *the proposition* that centaurs exist.[14]

But if the propositions which are invoked as *relata* for the relation of belief have the merit of providing true and false beliefs with *a common denominator* which they require, one may in rejoinder accuse them of obliterating the difference between truth and falsity, the relatum being as objective in the one case as in the other.

Now Moore feels that there is a serious objection here:

And here again I confess I can't put the objection in any clear and convincing way. But this is the sort of objection I feel. It is that, if you consider what happens when a man entertains a false belief, it doesn't seem as if his belief consisted merely in his having a relation to some object which certainly *is*. It seems rather as if the thing he was believing, the *object* of his belief, were just *the* fact which certainly is *not* — which certainly is not, because his belief is false.[15]

From this Moore draws two successive negative conclusions:

(a) "...if the object certainly is *not* — if there *is* no such thing, it is impossible for him or for anything else to have any kind of relation to it".[16]

In short, a relation without a relatum is not a relation.

(b) "And since there seems plainly no difference, in mere analysis, between false belief and true belief, we should have to say of all belief and supposition generally, that they *never* consist in a relation between the believer and something else which *is* what is believed".[17] In other words, if a false belief is not a relation, a true belief is not one either.

As one can see, Moore deliberately adopts the second of the solutions we indicated at the beginning. He refuses to construe belief as a relation.

What is the value of Moore's solution? It has an undeniable value. Contrary to Brentano's solution, it accounts for the *resemblances* between true and false beliefs. One may, furthermore, improve upon it so that it also accounts for the *differences* between true and false beliefs. One could, for example, explain the difference between them by saying that true beliefs are *behaviouristic properties* of the believer that, generally, have a greater chance of having favorable consequences for him than do false beliefs.

Despite its merits, Moore's solution remains defective. Its defect consists in being *incomplete*. It accounts for the resemblance between true and false beliefs, but undermines any hope of accounting for the resemblance between *true belief* and *knowledge*. If (true or false) belief is a property rather than a relation, it is irreparably cut off and detached from *knowledge*, at least as long as one does not also deny that knowledge is a *relation*. The latter denial would require a transformation of knowledge into a property of the knowing subject, which cannot be done without falling into the trap of idealism, something Moore is careful to avoid. Thus, Moore's program also only fulfills a part of the program that any satisfactory theory of belief and propositions ought to accomplish.

Some authors contest the legitimacy of this last requirement and maintain that there is no *resemblance* between *true belief* and *knowledge* which must be accounted for, but it is easy to show that they are wrong. Thus, for example, Cook Wilson thinks, as Robinson reminds us, that "...it is vain to seek such a common quality in belief, on the ground that the man who knows that A is B and the man who has that opinion both believe that A is B".[18] He invokes the following argument: "Belief is not knowledge and the man who knows does not believe at all what he knows; he knows it".[19]

It is clear that the author of *Statement and Inference* (1925) is impressed, and for good reason, by the difference in *epistemological status* that distinguishes knowledge from true belief. But this difference does not at all exclude the possibility of there being a resemblance in *logical structure* between knowledge and belief.

3. THE DISTINCTION BETWEEN PROPOSITIONAL VERBS AND COGNITIVE VERBS

A modern version of Moore's and Cook Wilson's solutions, and also just as incomplete, was proposed by Gale in 'On Believing what is not the Case' (1963), and it is taken up again in 'Propositions, Judgements, Sentences and Statements' (1967). Gale presents the outline of a theory which, in his opinion, accounts for false belief without the need to posit propositional entities.

It is therefore important that we examine this solution which compares favorably with nominalist conclusions upon which our investigation has touched thus far. Gale's solution, alas, is not acceptable. But his failure is instructive for, on the one hand, he shows clearly the insufficiency of expedient solutions and the necessity of fully elaborating a theory of the proposition, on the other, pointing up the particular defects in Gale's solution will provide us with the positive elements that will figure in our own solution.

Gale starts out from the fact that knowledge differs from belief in that it entails at least the reality of its object. He calls 'cognitive verbs' verbs expressing knowledge and 'propositional verbs' verbs expressing belief, and tries to show that these two categories of verbs cloak a *different logical syntax* under a *common grammatical syntax*.

> The differences which we have pointed out in the logical grammar of propositional and cognitive verbs enable us to give a very simple answer to the question of how we can judge, believe, and the like something when there is nothing to serve as the object of these acts. We judge or believe because propositional verbs, unlike cognitive verbs, are used in such a way that there need not be anything *in rebus* answering to their grammatical accusatives. How is it possible for a horse not to win the Kentucky Derby? Very simply, because we use the word 'horse' in such a way that what it designates need not be the winner of the Kentucky Derby. *That is the way we use these words*. There is no deeper mystery involved in the case of propositional verbs than in the case of horses if the above analysis is correct; for this reason there is no need to introduce Platonic complexes or anything else to serve as the object of propositional acts.[20]

Gale's expeditious solution is absolutely unacceptable. It is vitiated by an *ignoratio elenchi*. Gale demonstrates that the existence of a belief does not imply the existence of a *fact* as its object, but he concludes something different from what he has demonstrated; he concludes, in effect, that the existence of a belief does not imply the existence of a proposition as its object. Let us look at this in detail. It is Gale's intention to explain how we can believe when there is nothing to serve as an object of our belief. He then quietly abandons this question, which belongs to the *material mode of speech,* and answers another question which seems to be equivalent with the first, but is formulated in the *formal mode of speech*. Hereafter he answers the question how '*A* believes *p*' can be true without '*p*' being true, which is not the case for '*A* knows *p*'.

The answer to this question of logical grammar takes the from of a reasoning by analogy. Gale attempts to make us familiar with this contrast by comparing it with another well-known distinction between words designating tasks and words designating achievements. His reasoning may be summed up thus: '*A* knows *p*' implies '*p*' just as '*A* wins the race' implies '*A* crossed the finish line', and '*A* believes *p*' does not imply '*p*' just as '*A* is a competitor' does not imply '*A* is victorious'.

Unfortunately, Gale reintroduces the error of Moore and Cook Wilson. He explains only a part of the 'phenomenon', that is, the *disparity* between belief and knowledge. He leaves unexplained the *resemblance* between true belief and knowledge.

It is not his method that is defective, rather it is the way he employs it. By taking up again his tentative theory we may help to advance the debate, and to this end we must analyse very closely the logical syntax of propositional verbs.

4. THE LOGICAL SYNTAX OF PROPOSITIONAL VERBS

As a starting point for our comparison, take the words 'inventor' and 'researcher', rather than 'winner' and 'competitor'. '*A* is an inventor' implies '$(\exists x)(x$ is a discovery by $A)$', and '*A* is a researcher' does not imply '$(\exists x)(x$ is a discovery by $A)$'. If we now return to the contrast between knowledge and belief, we see that this must be formulated differently:

'*A* knows *p*' implies '*p* is true'.
'*A* believes *p*' does not imply '*p* is true'.

This comparison shows a *category difference* between the direct objects of these different transitive verbs 'to invent', 'to search', 'to know' and 'to believe'. Whereas the predicates 'true' and 'false' may be attributed to the direct objects of the last two verbs, they may not be *sensibly* attributed to the first two. For this reason we believe that one has to distinguish not between two verb categories, *cognitive* and *propositional*, but three: *cognitive* verbs ('to know'), *propositional* verbs ('to believe'), and intentional verbs ('to search').

But this is more far-reaching. The fact that 'A believes p' does not imply 'p is true' does not at all exclude that 'A believes p' may imply '$(\exists p)(p$ is believed by $A)$'. Now Gale's argument rests entirely on the hypothesis that this is excluded.

In order to dissipate the confusion that is at the bottom of Gale's peculiar supposition, we have to show in a more analytic way that the contrast between *to believe* and *to know* is not parallel to the contrast between *to search* and to *find*, and that the logical problems raised by the first two verbs, are not exactly the same as those raised by the last two.

This comparison will convince us that Gale's dichotomy (propositional verbs – cognitive verbs) must be replaced by a trichotomy (intentional verbs –propositional verbs – cognitive verbs).

The problem of *logical syntax* raised by the contrast between 'x is sought' and 'x is found' is the following: one may either give these verbs the *same* value range or a *different* one. If one gives them the *same* value range, then this will consist of *real* objects or *fictitious* objects endowed with a mentalistic existence. None of these three solutions is satisfying.

The first is unacceptable for the good reason that sought for objects do not always exist (the philosopher's stone, for example); the second solution is no more acceptable, but for the opposite reason: sought for objects sometimes exist. With respect to the second solution, which in substance amounts to assigning *categorically different* value ranges to 'to search' and 'to find', it forbids us, at the risk of an equivocal use of 'x', to formulate statements as indispensable as

'Peter found certain things he was looking for'.
$(\exists x)$(Peter was looking for x & Peter found x).

If, however, the value ranges differ only *numerically* (the objects found being a subset of the objects searched for just as the set of actual individuals is a subset of the set of possible individuals) the difficulty is avoided. This is perhaps a more correct way of construing Brentano's views.

However, when one examines more closely the contrast between *to believe* and *to know*, one sees that, *pace* Gale, it does not raise the same problems as that between *to search* and *to find*. While the difference between the object sought for and the object found has to do with *existence*, the difference between what is believed and what is known has to do with *truth*. The contrast is rendered in the phenomenologist terminology by the use of different words: *Objekt* for the object of research or discovery, *Objektive* for the object of belief or knowledge.

If we pay attention to this contrast, we shall be in a position to give the *same status* to the 'objektive' of belief and knowledge and we shall be able, at the same time, to account for the fact that belief can differ from knowledge in being false. Once it is acknowledged that the 'objektive' of belief belong to the *same* category, it is possible for the same variable to range *over the objects (objectives) of both knowledge and belief* without equivocation. The following sentences cease to cause trouble:

'Peter knows today certain things he only believed yesterday'
'($\exists p$)(Peter knows p today and Peter only believed p yesterday'.

It becomes possible to define knowledge as a subspecies of belief as in Ayer's *The Problem of Knowledge* (1956), where knowledge is defined in terms of 'true' and well-founded belief.[21] According to him, '($\exists p$)(p is known by A)' $=_{\mathrm{DF}}$ '($\exists p$)(p is believed by A) and (p is true) and (A has good reasons for believing p)'.

It also becomes possible to ascribe to *one* and the *same* entity these three roles: object (or objective) of knowledge, object of belief, bearer of a truth-value, and this is highly desirable indeed from a theoretical standpoint.

What sort of values shall we give to the variable 'p'? If we take these *values* to be sentences, we are committed to take *names* of sentences as substituends and to paraphrase:

John believes p and p is true

into

John believes 's' and 's' is true

One can even forge along Quine's lines the predicate 'believes-true'.

This solution is appealing. First it complies with one of the requirements imposed upon a 'theory' of propositions: the same sort of entity is taken to be the object of propositional attitudes and the bearer of a truth-value. Second it explains why materially equivalent or logically equivalent sentences cannot be substituted, in general, within belief contexts. Third it agrees with the view that the clause 'that s' is the nominalization of a sentence. This syntactic view is underpinned by several semantic arguments to the effect that the expression ⟨that s⟩ is a referring expression. In his unpublished dissertation *Propositions* (1976), R. Warner mentions these two inferences[22]

Smith believes *that Sally loves a man* therefore Smith believes *something*.

Smith believes that Sally loves a man and Jones believes *that Sally loves a man* therefore Smith believes that Sally loves a man and Jones believes *it* too.

and observes that both the existential generalization at work in the former example and the pronominalization at work in the second show that, *prima facie*, the expression *that Sally loves a man* is a referring expression.

There is, however, a difficulty which prevents us from taking up this nominalistic solution. One might object that mute animals such as dogs or cats are capable of entertaining propositional attitudes.

This being so, could we not take mental contents after all as *primary* bearers of truth, objects of propositional attitudes and values of the variable 'p' in the above-mentioned sentences? This is not a retreat to Platonism if propositions are conceived as mental entities. I shall even argue that such a solution can be said to be nominalistic in so far as the *mental entities* called upon *share* many essential features with *sentences*, a thesis I shall try to establish in the next section.

5. AN ATTEMPT AT ABSORBING PROPOSITIONS INTO SENTENCES

Ryle begins by establishing that the proposition, which is the presumed object of belief, must have a *syntactic* structure analogous to that of the fact, which differs from a collection of objects like a sentence differs from a list. Ryle is then interested in showing that in order to account for propositions' capacity of being true or false, one must conceive of them not as self-sufficient entities, but as entities pointing to something other than themselves:

If a proposition is one entity and if it corresponds to or is similar in structure to a fact, why should it be called true of that fact, and not merely *like* it or *analogous* to it? ... What is the relation between (a) a true proposition and the fact that it is true of and (b) a false proposition and the fact or facts that it is false of? Another way would be to ask, how, if propositions are independent of other substances, are they *about* them?
...in the case of false propositions we should have to grant not merely that it was unlike some given fact but that it was unlike *any* fact. So we could not speak of *the fact* that it was false of; and consequently we could not speak of its being false *of* anything at all.[23]

Under the pressure of common problems, British and German philosophy has a parallel development. The transition of the proposition as a substantial entity that Moore defends in 'The Nature of Judgement' (1889), to the proposition conceived as pointing to something else, is paralleled by the transformation one notes between Husserl's conception of meanings in *Logische Untersuchungen* (1900–1901) and that in *Ideen zu einer reinen Phänomenologie* (1913).

In a long note inserted into the French translation of *Ideen*, Ricoeur sums up the innovations contained in this treatise on the point which is of interest to us:

> Husserl here declares that the most intimate part of intentionality has not been accounted for if, in the 'sense' aimed at, by the consciousness, one does not discern in addition an arrow which goes through the sense and indicates the direction towards... or the claim to... reality [objectivité].[24]

Let us now return to Ryle's text to extricate the line of argument it contains. By asking rhetorical questions, Ryle wants to make us aware of something like the following: when we say that a proposition is true if it effectively *corresponds* to a fact, we are speaking elliptically. To be complete we should say: A proposition is true if it corresponds to the fact which it purports to correspond to *by virtue of its meaning*. In other words, Ryle wants us to understand that one cannot account for truth by a simple correspondence relation between a proposition and a fact, nor for falsity by a simple absence of such a correspondence. In his view, to claim the contrary would be to ignore the defective nature of *falsity*, to which we have already alluded, a trait that neither *absence* nor *negation* possess. A false proposition, as opposed to a negative proposition, is a *failed attempt* at correspondence, rather than a pure and simple absence of correspondence. To account for this failed attempt one must admit a banal truth often overlooked, namely, that the proposition is not a *self-sufficient* being, but rather something that is directed toward something other than itself, or, to use Ryle's own words, is *about* other substances. In other words, we must admit that the proposition *signifies* or that it *has a sense*.

It is possible today to develop and improve upon Ryle's argument in a way which justifies not only the rejection of propositions but also *their replacement* by sentences and to complete thereby the semantic conversion of the word 'proposition'. To extend Ryle's argument, it suffices to think of the following fact: one cannot attribute *possession* of a sense to propositions. Generally, one says that words *have* a sense, that sentences *have* a sense, that statements *have* sense; one also sometimes says that propositions *are* the sense of sentences or statements, but one never says that propositions themselves *have* sense, for this would lead to an infinite regress.

Thus, when one wants the proposition to play the role of the object of belief, the object of knowledge, and the subject of the predicates 'true' and 'false', one is forced to attribute a meaning to it and to confer on it all the prerogatives of the sentence.

6. SEARLE'S VIEWS ON INTENTIONALITY

Searle's essays on 'Intentionality and Use of Language' and 'What is an Intentional State' contain views which are quite in line with that developed by me in the former section. However, before claiming that Searle's arguments lend support to my thesis, I should summarize them.

Confronted with the task of accounting for the fact that a intentional state such as belief can be *about* a non-existent object (Pegasus) or an inexisting state of affairs (Pegasus' flight), Searle offers the following answer: "Intentional states represent objects and states of affairs in exactly the same sense that speech acts represent objects and state of affairs".[25]

In other words, a state of belief is not aimed at a metaphysical entity such as proposition — that which prompted Wittgenstein's exclamation:

'A proposition is a queer thing!' Here we have in germ the subliming of our whole account of logic. The tendency to assume a pure intermediary between the propositional *signs* and the facts.[26]

On the contrary, Searle claims that on his account:

...there is nothing ontologically peculiar about intentional objects; they are just ordinary objects and states of affairs at which our mental states are directed. To say that Sally is the intentional object of Bill's love of Sally is just like saying Sally is the described object of Bill's description of Sally, it assigns no ontological peculiarity to Sally.[27]

Searle's position is thus in perfect agreement with Husserl's thesis as it is understood by Ricoeur in the latter's quotation (p. 165).

I think that one could sum up Searle's account of intentionality by saying that what is "queer" in the relation inherent to intentional states is not the *relatum*, the *object*, the *target*, but the relation itself which is nothing but the relation of representation exemplified by sentences and pictures.

There are even more likenesses between sentences and propositions than between sentences and pictures. Although "propositions", as Stalnaker observes, "have no syntax, no 'exact words' or word order, no subjects, predicates, or adverbial phrases", they "can stand in logical relations like implication, independence and incompatibility".[28] This is true of propositions considered either as *platonistic entities* or as *intentional states* like beliefs. The point has been made by Harman in *Thought* (1974):

Consider logical relations between beliefs. If the belief that *snow is white and grass is green* is true, the belief that *snow is white* is true ... and so forth. Clearly, a generalization is possible: whenever the belief that *P & Q* is true, the corresponding belief that *P* is true. Some such general statement is appropriate. We cannot simply list the relevant instances, since there are an infinite number of them. But the relevant generalization presupposes that certain beliefs have the structure of conjunction.[29]

By stressing the fact that beliefs or other intentional states are *like* sentences in having a representative relation with the world and in being made out of parts, Searle does not want to reduce intentional states to sentences: "By explaining Intentionality in terms of linguistic acts, I do not mean to suggest that Intentionality in somehow essentially linguistic" and he goes on to say, "Language is derived from Intentionality, and not conversely".[30]

Searle's rejection of a derivation of Intentionality from language is sound. It is generally agreed that belief exists among mute animals such as cats and dogs. Equally, the scope of intentional states is wider than can be linguistically expressed. One can believe and even know that all the propositions asserting an identity between a real number and itself are true, but since the set of real numbers is not denumerable, no recursively enumerable language is rich enough to offer a sentence corresponding to all these "facts" — I use "fact" in an inverted commas sense — about real numbers.

The only point where one might disagree with Searle is when he claims that "ground floor intentional states" (beliefs about the world as opposed to beliefs about beliefs) are extensional: "Just as my statement that Caesar was Emperor of Rome is extensional, so my belief that Caesar was Emperor of Rome is extensional and for the same reasons".[31] Searle's claim is hardly tenable in the light of Church's argument quoted Chap. IX, Sect. 4.

REFERENCES

[1] B. Russell, 'On Propositions: What They Are and How They Mean', (1919) reprinted in R. Marsh (ed.), *Logic and Knowledge*, Allen and Unwin, 1956, p. 285.
[2] F. Brentano, 'The Distinction between Mental and Physical Phenomena', transl. by D. B. Terrell and R. Chisholm, in R. Chisholm (ed.), *Realism and the Background of Phenomenology*, Glencoe, Free Press, 1960, p. 51.
[3] Plato, *Theatetus* (189A), London, William Heinemann 1921, p. 175.
[4] Malebranche, quoted by Mgr. Dies, Palto, *Théétète*, Les Belles Lettres, Paris, 1963, p. 188.
[5] R. R. Blake, 'On McTaggart's Criticism of Propositions', *Mind*, (1928) 448.
[6] G. E. Moore, *Some Main Problems in Philosophy*, Allen and Unwin, London, 1953, p. 57.
[7] Moore, *Ibid.*, p. 58.

⁸ Moore, *Ibid.*, p. 62.
⁹ Moore, *Ibid.*, p. 63.
¹⁰ J. Ruytinx, 'Logique leibnizienne et philosophie contemporaine', *Proceedings of the XIIIe International Congress of Philosophy*, Vol. V, Mexico, 1964, p. 320.
¹¹ G. Ryle, 'Are there Propositions?', *Proceedings of the Aristotelian Society*, 1929–1930, p. 107.
¹² Brentano, *Op. cit.*, p. 50.
¹³ Moore, *Op. cit.*, p. 259.
¹⁴ Moore, *Ibid.*, p. 259.
¹⁵ Moore, *Ibid.*, p 263.
¹⁶ Moore, *Ibid.*, p. 263.
¹⁷ Moore, *Ibid.*, p. 265.
¹⁸ R. Robinson, 'Cook Wilson's View of Judgment', *Mind*, 37, (1928) 344.
¹⁹ Robinson, *Ibid.*, p. 344.
²⁰ R. Gale, 'Propositions, Judgments, Sentences and Statements', *Encyclopedia of Philosophy*, Vol. 6, p. 500.
²¹ A. J. Ayer, *The Problem of Knowledge*, Pelican, London, 1956, p. 35.
²² R. Warner, *Propositions*, (unpublished 1976).
²³ Ryle, *Op. cit.*, pp. 108–109.
²⁴ P. Ricoeur, in Husserl, *Idées directrices pour une phénoménologie*, transl. by P. Ricoeur, Gallimard, Paris, 1950, pp. 431–432.
²⁵ J. Searle, 'What is an Intentional State?', L.A.U.T., Trier, March 1978, Serie A, Paper No. 50, p. 14; and in *Mind* 88, (1979) 75.
²⁶ L. Wittgenstein, *Philosophical Investigations* § 94, Transl. by G. E. M. Anscombe Oxford, Blackwell 1953, p. 44.
²⁷ J. Searle, 'Intentionality and Use of Language', L. A. U. T., Trier, 1978, p. 7.
²⁸ R. Stalnaker, 'Propositions', in A. F. Mackay and D. D. Merrill (Eds.), *Issues in the Philosophy of Language*, Yale University Press, 1976, p. 83.
²⁹ G. Harman, *Thought,* Princeton Univ. Press, 1974, p. 55.
³⁰ J. Searle, 'What is an Intentional State?', *Op. cit.*, p. 14. *Mind*, p. 75.
³¹ J. Searle, 'Intentionality and Use of Language', *Ibid.*, p. 6.

CHAPTER VII

PROPOSITIONS AS MEANINGS OF SENTENCES

1. THE RELATIONAL CONCEPTION OF MEANING

In the preceding chapter I approached propositions by way of philosophical psychology, defining propositions as the objects of belief. But a thorough analysis led us to lend credit to the view that 'the subject x believes the proposition p' can be analyzed as 'x is ready to assent to the sentence s which *means* that p' or 'x entertains a belief-content b which *represents p*'.

In *Thinking and Meaning* (1947), Ayer writes:

> For we are now in a position to see that the current analysis of knowing, believing, doubting, judging, imagining; and all the other modes of thought, as acts of the mind which are directed on an object, whether it be a fact or a proposition, or what you will, is largely mythological. For the only way in which an 'object' comes into the picture at all is as the referent of a sentence.[1]

I can to a certain extent subscribe to this view since I rejected the analysis which construes belief as a dyadic relation holding between a mind and a propositional content and by the same token the analysis which construes *knowledge* as a binary relation holding between a mind and a fact. In the two cases I proposed instead a triadic relation holding between a mind, a sentence or a propositional content and what they 'represent'. It is now incumbent upon me to examine what this representing ingredient consists in. The risk exists that propositions crop up again in a new role: as *meanings* of sentences or as *relata* of the relation expressed by the verb 'to mean'.

That this risk is not a figment of our imagination can be proved easily. Consider the following passage of Ayer's: 'Meaning and Intentionality' (1958): "They [the propositions] are the vehicles of truth and falsehood, the objects of the various cognitive attitudes *and the meanings of the sentences in which these cognitive attitudes may be expressed*".[2]

At first sight this statement seems to be a very clear and concise formulation of the unified theory of meaning I am trying to build up. Yet I cannot agree with it without qualification. If I take as I do, propositions to be intentional states *representing* reality on a par with sentences or pictures, I am not allowed to say that they are *represented* or *meant* by sentences.

In other words I have to look for a theory of meaning which fits in with the theory of intentionality I took up in the preceding chapter. Roughly speaking, my position is this: Intentional states such as belief-contents represent reality, to borrow Searle's word, in the same way as sentences. When I stress the *similarity* between propositions understood as belief-contents and sentences, however, I understand the word 'sentences' as denoting linguistic signs *endowed with meaning* which give them their representative character. Those who claim that propositions are needed as the meaning of sentences use the word 'sentence' in a different way, i.e. as denoting a string of words taken as *phonetic* and *syntactic* entities as opposed to *semantic* entities.

We ought to be aware of this *equivocation*, but we should refrain from thinking that nothing more is involved than a *terminological disagreement*. As a matter of fact, those who construe meaning as a relation holding between signs taken as strings of marks or sequences of sounds and significata, also construe significata as 'queer entities' like Platonistic entities or Meinongian entities. I shall examine a clear case of that kind of hypostatization in the next section.

2. THE ETERNALITY AND TEMPORALITY OF MEANING

In Husserl one finds the purest example of the conception of the proposition as the second term of the relation *to signify*. A radicalized version of Husserl's views was defended by M. Combes in *Le concept de concept formel* (1970).

In *Logical Investigations*, Combes recalls, Husserl affirms that

> Meaning is related to varied acts of meaning — Logical Presentation to presentative acts, Logical Judgment to acts of judging, Logical Syllogism to acts of syllogism — just as Redness *in specie* is to the slips of paper which lie here, and which all 'have' the same redness. Each slip has, in addition to other constitutive aspects (extension, form, etc...), its own individual redness ... .[3]

Husserl's conception, according to which the noema is immanent to the noesis, does not, however, sufficiently separate the thought object from the thinking subject, according to Combes who writes:

> One can dispute in every instance this identification of the sense of a judgement as '2 + 2 = 4' or the theorem about the sum of the angles of a triangle with what the corresponding acts of thought have in common. In fact, these judgements are what acts of thought are aimed at. Now if it is true that acts of thought can have in common the fact of aiming at the same object, this aiming remains always outside the object itself. To be aimed at is no part of the essence of the object aimed at ... . How else could one distinguish the senses of '2 + 2 = 4' from that of '2 + 2 = 4 is thought' and that of 'That 2 + 2 = 4 is thought is thought'?[4]

Confident of his arguments, Combes fearlessly follows a Platonistic conception of meanings. "... are meanings not ideal objects which eternally subsist whether we think of them or not? ... In a sense yes".[5]

However, Combes considers that, as Platonistic as it is, his doctrine is not exposed to the same objections as Plato's theory of Ideas, for it differs from it in an important respect: his 'Meanings' are not *individual* things.

But, apart from this, these meanings have all the features of Plato's Ideas, for example, *immutability* and *independence with respect to the thinking subject*. Combes even presents the following as theorems:

(1) 'Meanings evolve' is devoid of sense;

(2) 'Meanings are produced' is devoid of sense;

(3) One cannot regard meanings as products of the thinking that thinks them.

It is obvious that the conception of meaning contained in these statements is in contradiction to the nominalism we have defended thus far. Therefore, we must ask ourselves whether Combes' *arguments* suffice to support this conception and whether the facts support it. We believe that we can prove the contrary.

In order to establish the independence of a *judgement* with respect to the *judgmental activity*, that is, of the proposition with respect to the *activity of thinking*, Combes reasons by analogy: "To be aimed at is no part of the essence of the object aimed at". But to thus assimilate the proposition to the autonomous *object, existing prior* to the aiming, rather than to the *objective*, that is, to the object determined by the shooter's thought as the point of aim, is for Combes to grant himself precisely what he wishes to prove. His argument is, therefore, vitiated by begging the question.

On the other hand, it is not true, if one defines propositions as products of the act of thinking, that one can no longer distinguish the sense of '$2 + 2 = 4$' from that of '$2 + 2 = 4$ is thought', and from that of 'that $2 + 2 = 4$ is thought is thought'. Why would the product have to be conceived as preserving a trace of its origin? The nominalist holds that the proposition depends *causally* and *ontologically* on an act of thought, but he does not claim that it takes itself as *subject-matter*, and he is, besides, not at all constrained to do so.

In other words, the nominalist affirms that the existence of '$2 + 2 = 4$' implies the existence of an act of thought yielding '$2 + 2 = 4$'. But this affirmation does not oblige him to hold that '$2 + 2 = 4$' implies 'the thought that $2 + 2 = 4$ exists'. One might be forced to admit, however, that a serious utterance of 'S' pragmatically implies the belief that 'S' is true. " $2 + 2 = 4$ and nobody believes it" is pragmatically inconsistent in the same way as

" '2 + 2 = 4' and I don't believe it". This is not, however, an issue on which the nominalist is committed to more idealism than common sense allows. A careful distinction between sentence and utterance, logical implication and pragmatic implication, enables him to account for the fact that '2 + 2 = 4' was true before man came into existence.

We have just examined the arguments put forth by Combes in support of his thesis. We shall now show that the *thesis* itself does not stand up to examination. Combes is well aware of the fact that science has a history, and that the idea mathematicians have today of space is no longer like Euclid's, but he does not see in this a counter-example to his conception of meanings: "... The point is to know if meanings really evolve or if we discover new meanings and forget the old".[6] It is the second description of facts which he adopts.

If the two descriptions accounted *equally* for the complexity of the facts, we could hold nothing against Combes. But this is not the case, far from it. We shall see that if one adopts Combes' suggestions, one deprives the lexicologist of the possibility of speaking of an 'extended sense' or 'a *shift* in sense through metaphorical usage'.

On the other hand, if one treats meanings as things which undergo change through time, as has been done by lexicologists since Herder, one will be able — as L. J. Cohen remarks in *The Diversity of Meaning* (1962) — to "... discuss, as Herder's predecessors could not, how a decline in the relative importance of pastoral wealth extended the meaning of the Latin word 'pecunia', or how the use of the rosary gradually changed that of the English word 'bede' or 'bead' ".[7] Cohen's example well illustrates the scientific impoverishment that would result from the adoption of Combes' views. Were they adopted, one would in fact have to present the resemblance between the word *pecus* and the word *pecunia* as a fortuitous homonymy. But the other description is much more helpful.

From all of this it is clear that we must reject Husserl's and Frege's relational theory of meaning and the Platonistic conception of propositions which results from it. But what are we to put in its place? In order to answer this question we shall examine proposed replacement theories so as to find, if not a complete answer to our questions, at least some material from which we might construct a theory that does not yet exist.

3. THE BEHAVIOURISTIC ANALYSIS OF THE MEANING OF SENTENCES

Anxious to reconcile a realist conception of truth with a nominalistic conception of sentence-meaning, Ayer proposed in 'Meaning and Intentionality'

(1958), which we mentioned above, an extremely ingenious theory of the meaning of sentences. He begins with the notion that a sign tends to elicit the same response as that which it signifies, and he seeks to understand the mechanism of meaning by applying the model of the conditioned reflex in animals, as did Morris before him. "...The guiding principle is that a sign tends to evoke the same behaviour as that which it signifies. The model used is that of the conditioned reflex in animals or one's normal reaction to a fire alarm."[8]

The model of conditioned reflex is, at first blush, promising. It seems to present *in simplified form* a resolution of the enigma of the meaning of a sentence, a meaning which, we repeat, must be the *same* whether the sentence is *true* or *false*.

We must, in fact, recall the circumstances of salivation conditioning in dogs. The conditioning is produced by the repeated association of a sound signal with the presentation of food. When the reflex is induced, the production of the signal *by itself* provokes salivation, *even if it is not followed by food*. This fact deserves some consideration: the signal has the *same* meaning whether the event it announces is produced or not. The proof is that one may deceive the animal. The signal is thus *two-valued* like the sentence.

Ayer, however, proposes a more elaborate theory than Morris. He does not merely assimilate false sentences to misleading signals, for he is aware of the fact that false sentences do not always release reactions in the hearer. Indeed, when the hearer does not *believe* the false sentence he does not react. In this light, Ayer believes that we must perfect the behaviouristic theory of meaning in such a way that it accounts for this fact.

The definition of meaning which he proposes, therefore, contains a clause relating to belief:

... the meaning of indicative sentences can be analysed as follows: given that S is a sentence, p a proposition, and A a person, S means p to A if and only if A's assenting to S is constitutive of his believing p. It is not of course implied that A does actually believe that p, all that is required is that if he did believe it his assent to S would be constitutive of the belief.[9]

For example, 'lions are carnivores' *means* for A that lions are carnivores if A is *disposed to give his assent* to this sentence if he believes that lions are carnivores. *Meaning* is thus reduced to a *disposition* to assent to a sentence, a disposition which is the same whether the sentence is true or false. *It is, therefore, tempting to say that meaning is nothing else but a disposition to respond.* With respect to belief, it is ultimately also analysed in terms of behavior. (to believe that p is to act in a way that is appropriate if and only if p).

Ayer's theory contains several important insights. First, by giving a *dispositional* analysis of meaning, Ayer provides us with a theory of *semantic competence*. This is no mean achievement. The need for a theory of semantic competence as opposed to semantic *performance* was not felt at that time as can be seen in the following statement made by the linguist Rulon Wells only four years before Ayer's paper. In 'Meaning and Use' (1954), Wells seems to be unaware of the distinction between competence and performance when he writes:

The most obvious way to conceive meaning is to conceive it as a dyadic relation between a sign and an object.... A rival account... recognizes three factors: sign, object, and user. But a moment's thought will prove that even this triadic conception is not exact. For instance, we would need to mention the time of signification: A is a sign of B to C at time D;...[10]

A second merit of Ayer's theory is this: It is not the objects of meaning or belief relation which are queer but the relations themselves. The relational terms 'means that', 'believes that' are paraphrased away in behavioral terms.

A third merit of Ayer's account lies in its ability to explain why sentences can have the same meaning, whether they are true or false.

Unfortunately, Ayer's solution to the problem of analyzing belief which is a part of his analysis of meaning is hampered by a serious objection. If we define 'A believes that p' as

'$(\exists x)$. A is disposed to conduct himself in the manner x.

[x is a behavior appropriate for subject A] $\equiv p$

then, as Church[11] has shown in 'Logic and Analysis' (1958) the 'p' in the analysans as well as the equivalence '$\equiv$' being extensional, one may substitute for this sentence any other sentence with the same truth-value. It then follows from Ayer's analysis that if a person A has one false belief, he has every false belief. And if he entertains one true belief, he has every true belief.

4. THE CHESS-THEORY OF MEANING

Another manner of dispensing with the relational account of meaning which leads to the hypostatization of queer entities like proposition can be found in Ryle's essay 'Theory of Meaning' (1956).

In this essay, Ryle launches a general attack against the conception according to which meaning is a two-termed relation linking expressions with extra-linguistic entities. According to him, the defenders of this theory make an absolutely unacceptable generalization: they assimilate all meaning to denotation. In other words, they generalize to all linguistic expressions the type of semantic relation that exists between a proper name (the name 'Fido', for example) and the individual bearing that name (the dog Fido). He called this the 'Fido'-Fido theory. Notice that Ryle's simplified picture does not fit theories like that of Charles Morris which distinguish between *designatum* and *denotatum* (a designatum which actually exists).

As a substitute solution Ryle offers a version of Wittgenstein's sense-use theory. In his opinion, this theory should not complete or extend the 'Fido'-Fido theory but, on the contrary, should radically eliminate it. The following passage clearly indicates that Ryle considers the denotational theory to be completely incompatible with the sense-use theory:

If the meaning of an expression is not an entity denoted by it, but a style of operation performed with it, not a nominee but a rôle, then it is not only repellent but positively misleading to speak as if there existed a Third Realm whose denizens are Meanings.[12]

Ryle then establishes this in two ways. First by way of an analogy from chess shared with Saussure and Wittgenstein, and then with the help of the analogy of syllables from Plato:

We can distinguish this knight, as a piece of ivory, from the part it or any proxy for it may play in a game of chess; but the part it may play is not an extra entity, made of some mysterious non-ivory. There is not one box housing the ivory chessmen and another queerer box housing their functions in chess games. Similarly we can distinguish an expression as a set of syllables from its employment. A quite different set of syllables may have the same employment. But its use or sense is not an additional substance or subject of predication. It is not a non-physical, non-mental object – but not because it is either a physical or a mental object, but because it is not an object.[13]

Ryle's virulent attack against theories of meaning which uniformly reduce all meaning to denotation has much less scope than is generally believed. It cautions us to pay attention to the fact that not all parts of speech play the same role. But what are the pertinent *differences in role*? This depends on the discipline.

There may be good reasons in grammar for distinguishing the role of the subject from that of the predicate, and the logician may have equally good reasons for neglecting this distinction, because he has other objectives in mind than those of the grammarian. The logician's analyses are guided by his preoccupation with formulating rules of valid reasoning. For example, in order to account for the validity of the syllogism, which must admit substitutions for the middle term, we must adhere to the rule "... according to which each concept (general term) must be able to be subject or predicate without changing its meaning".[14]

It can also be objected that the rejection of the relational theory of meaning as conceived by Ryle is much too radical a measure. As we have already seen, there are signs in language for which the relational theory of meaning is fully valid. They are quantified variables and proper names.

The scope of the denotation relation has shrunk, but the verb 'to denote' remains a *dyadic* predicate expressing a relation, although the relata of this relation are presently limited only to values of bound variables. Thus, there is a domain from which the relational conception of meaning is unexpungeable.

It is well-known that Russell himself restricted the luxuriant ontology of the *Principles* in 'On Denoting' (1905), and it is thanks to the theory of descriptions that we have the first serious restriction imposed on the 'Fido'-Fido theory. The polyadic theory of judgement defended in *Problems of Philosophy* (1911) and *Principia Mathematica* (1st edn., 1913) is another.

These first steps toward emancipation still appear too weak to Ryle who reproaches Russell for remaining too attached to the 'Fido'-Fido theory. "But he [Russell] was, I think, still held up by the idea that saying is itself just another variety of naming, i.e. naming a complex or an 'objective' or a proposition or a fact − some sort of postulated *Fido rationis*".[15]

In 'Use, Usage and Meaning' (1961), Ryle insists again on the difference between sentence's meaning and words' meaning and attacks Wittgenstein and Strawson for ignoring it:

> Some philosophers, oblivious of the distinction between Language and Speech,... give to sentences the kind of treatment that they give to words, and, in particular, assimilate their accounts of what a sentence means to their accounts of what a word means.[16]

Ryle is undoubtedly right in stressing the difference between sentences and names. Sentences are not *names* of queer entities. But from this it does not follow that sentences are *not related* to reality. It follows only that the semantic relation at hand, if such there be, is not a referring relation. What else is it?

5. AN ATTEMPT AT DISSOLVING THE PROBLEM RAISED BY THE MEANING OF SENTENCES

Ryle tried to give a positive answer but did not succeed. In order to take account of the difference between sentences and words, he invokes Saussure's distinction between *langue* and *parole*:

> We are tempted to treat the relation between sentences and words as akin to the relation between faggots and sticks. But this is entirely wrong. Words, constructions, etc., are the atoms of a Language; sentences are the units of Speech.[17]

It is interesting to compare Ryle's text with that of Benveniste's, which is rigorously contemporary:

Sentences, which can be newly created indefinitely and are of limitless variety, make up the life-blood of language in action. ... With sentences one leaves the domain of language as a system of signs and enters into another universe, that of language as a tool of communication, the expression of which is discourse. It is by this means that one can define it: the sentence is the unit of discourse.[18]

Ryle's and Benveniste's attempts to account for the difference between sentences and words fail. In both cases an important distinction is lacking, the distinction between *sentence-type* and *sentence-token*. Even if we recognize that sentence-tokens are the units of speech, we are not for that matter freed from the obligation of accounting for the meaning of sentence-types which belong to language as opposed to speech just as well as words.

If we denied use and therefore meaning to sentences on the ground of Ryle's argument we would quickly run into paradox. In *Grundlagen der Arithmetik* (1884), Frege writes: "... it is only in the context of propositions that words have any meaning...".[19] If we accept Frege's thesis and, at the same time, the thesis according to which a sentence is not the sort of thing which has denotational meaning, we would be brought to the absurd consequence that *neither sentences nor words* have meaning. Those who accept Frege's disputable thesis have to conclude that Ryle has not succeeded in explaining how sentences have meaning nor has he explained how that very meaning *relates* linguistic entities to the world. Rather than solving the problem he has denied it.

6. THE PICTURE THEORY OF MEANING

Wittgenstein anticipated Ryle's attack against the 'Fido'-Fido theory applied to sentences in the following passage from the *Notebooks* (1914—16):

Frege said 'propositions are names': Russell said 'propositions correspond to complexes'. Both are false; and especially false is the statement 'propositions are names of complexes'. Facts cannot be named. The false assumption that propositions are names leads us to believe there must be 'logical objects': for the meaning of logical propositions would have to be such things.[20]

Wittgenstein, however, did not give up the ambition to offer a relational treatment of the sentence meaning. One can sum up his view on the matter as it is expressed in the *Tractatus* in this way: sentences do not *name* states of affairs, they *depict* them in the way a picture depicts reality: "A proposition can be true or false only in virtue of being a picture of reality".

This view is a decisive step toward accounting for the queerness of the representative virtue of sentences and intentional states pointed out in the preceding chapter along the lines of Ricoeur and Searle.

The picture theory has several merits. First it clearly distinguishes not only between *sentences* and *names* but also between *sentences* and *lists of names* whereas alternative theories have been criticized for confusing them. For instance, Ryle holds that Carnap was guilty of the second confusion. Commenting on the following passage: "Any proposition must be regarded as a complex entity, consisting of component entities, which, in their turn, may be simple or again complex. Even if we assume that the ultimate components of a proposition must be exemplified, the whole complex, the proposition itself, need not be".[21] Ryle wrote: "A sentence is, therefore, after all, just a list. 'Socrates is stupid' is equivalent to 'Socrates, attribution, stupidity'. ...Plato knew better than this...".[22] Gale takes over this critique in 'Propositions, Judgements, Sentences and Statements' (1967), as follows:

> The difficulty with this solution, ... is that it assimilates saying to referring ... and is therefore unable, in the same way as was the multiple-relation theory, to account for the peculiar internal unity of the sentence used to express a proposition. To say 'Desdemona loves Cassio' is not just to mention three things – Desdemona, loves, and Cassio; that would be a mere laundry list.[23]

A second merit of the picture theory of meaning is that it accounts for an essential property of sentences which was brought to the foreground of linguistic research by Chomsky and his followers, namely the immediate intelligibility of a sentence whose words are understood: ("It belongs to the essence of a proposition that it should communicate a *new* sense to us") *Tractatus logico-philosophicus* 4.027).[24]

That the picture theory of meaning explains this feature of sentences is easy to show. Pictures are easily intelligible. We do not require a caption under a sketch representing a car accident to explain what the drawing represents. The reason for this transparency, for this intelligibility of pictures, lies in the fact that the *same* key allows us to decipher *all* pictures: the principle of isomorphism. Thus, one may learn, once and for all, this key, this convention. We definitely have to do here with a *convention*. The resemblance between the original and a copy may well be very great, but, as is well known, this is not enough to turn the *copy* into a *representation*. Two drops of water may *resemble* each other. It does not follow that one of them *represents* the other. Only a convention can achieve that goal.

In the reproduction of the picture there is a categorical homogeneity between the picture and what it represents: individuals are represented by individuals, relations by relations, and facts by facts. In an atlas, which is already more sophisticated than a picture, lines represent rivers and the relations of distance between these lines represent the relations of distance between rivers. This is one of the requirements of isomorphism. The *fact* that a line is twice as long as another represents the fact that a river is twice as long as another. If one takes the claim according to which propositions are pictures literally, one will have to account for the categorical homogeneity between the linguistic sign and the thing signified: the sign for an individual will necessarily be an individual, the sign for a relation a relation, and the sign for a fact a fact.

Without hedging, Wittgenstein assumed these consequences of the picture model. In 3.14 of the *Tractatus* he writes: "A propositional sign is a fact", and in 3.1432: "Instead of, 'The complex sign "*aRb*" says that *a* stands to *b* in the relation *R*', we ought to put, '*That* "*a*" stands to "*b*" in a certain relation says *that aRb*'.[25]

It stands out in this passage that for Wittgenstein the sentence is not a list of names of individuals and of *names of relations*, but a sequence of names linked by an objective relation: "An elementary proposition consists of names. It is a nexus, a concatenation, of names". This conception readily accounts for the cohesion of the sentence. It further has the merit of accounting for the way in which a syntactic factor, the word order, contributes to the meaning of a sentence.

Suppose that the relation *greater than* that exists between the numbers 5 and 3 is represented by the relation *to the left of* that exists between the names of the numbers, '5' and '3'. One sees immediately that it is convenient, within the framework of this symbolism, to account for at least one of the differences in the sense that is unaccounted for by the conception of a sentence as an unordered collection of words which Ryle attributes — perhaps wrongly — to Carnap. We must only think of the difference in sense produced by permuting the terms in a sentence containing a predicate expressing an asymmetric relation.

7. THE LIMITATIONS OF THE PICTURE THEORY OF MEANING

The *assimilation* of sentences to pictures does not only carry advantages, it also raises difficulties, a close examination of which is instructive for our purpose. A delicate question immediately arises for the exegeticist as well as for the philosopher intent upon saving what is valid in this theory: how far exactly can one push the analogy between sentences and pictures, either in order to be faithful to Wittgenstein or to take account of facts? In any event, it is certain that Wittgenstein did not conceive of sentences as ideograms. He obviously did not want to say, when speaking of pictures, that true sentences *physically* resembled the facts described, as onomatopoetics resemble the sounds they describe.

As J. Morrison says in *Meaning and Truth — Wittgenstein's Tractatus* (1968),

... the proposition-picture and what is pictured (the fact) need not be of the same *kind*. For instance, a spatial relation in the sign can picture a temporal relation in the fact, as when musical notes in the printed score are arranged spatially, but when played or sung are temporal. The score, nevertheless, remains a picture of the music as it is actually performed".[26]

Thus, the way in which the relationship of a sentence to a word resembles the relationship of a picture to an object is not physical, but logical or, if one wishes, categorial. Signs belong to the same or to a different category depending on whether their *meaning* does. Just as in a picture of a cat on a mat, the relation *being on* categorially differs from both the cat and the mat, in a sentence the words categorially differ from the relation between them.

This interpretation seems to interpret faithfully Wittgenstein's thought, and to bring out a peculiar consequence of the picture theory conceived as a model for the meaning of sentences, a consequence pointed out by J. Rosenberg in "Wittgenstein's Theory of Language as Picture' (1968). In pictures all pencil lines or all color spots belong to the same category; they are individuals. In a sentence, however, all the words belong to the same category only if one abstracts from their grammatical function. In fact, if one takes the latter into consideration, one ought to assign the word '(...) between (...)' to the category of relations and the word 'Socrates' to the category of individuals.[27]

If Wittgenstein takes the picture model seriously, he must assign all words to the same category. Now, what he does — and this confirms our interpretation — is to reduce all *words* of a sentence to *names*. At first blush one might think that not much harm is thus done to the specificity of the sentence. The *difference in function* between names and predicates that is erased by reducing all the words of a sentence to names seems, in fact, to be taken over by the *categorial difference* between words that are individuals and the syntactic relation of concatenation which is a relation.

However, looking at the matter more closely, one notes that this takeover is not so easily achieved. In a picture, in effect, there are innumerable spatial relations. In a sentence, on the other hand, due to its one-dimensionality, there is only one: *being to the right of* or its converse, that is, the relation of concatenation.

8. BEYOND THE PICTURE THEORY

The picture theory contains important insights which any theory of the meaning of sentences will have to retain. The conception of sentences as a *structured* complex is a case in point. The main weakness of the picture theory lies in too *narrow* a conception of structure.

It is true that the *arrangement* of the words in a sentence contributes to the meaning of the sentence, but they do so in a much *more sophisticated* way than the arrangement of white and black lines on a T.V. screen contribute to the picture being what it is, i.e. being a picture of so and so.

Admittedly, in some privileged cases *syntactic structure* reduces to *iconic structure*.

It is not necessary to go very deeply into grammar to realize that the mode of representation of natural language shares several traits with picturing.

The semantic role of word order is a case in point. Consider the following two sentences taken from Vandamme's *Simulation of Natural Language* (1972):

(1)   Evy is killed by Mike.
(2)   Mike is killed by Evy.

As Vandamme observes, "Although the two sentences have identical morphemes and consequently also the same significata as sub-parts, the significatum of these two sentences are different. The word-order in the sentences put the same significata in different relations to each other. You can even assert that the difference in the significatum of sentences (1) and (2) is caused by the different position which is taken by the same significata".[28]

The use of *trees* for removing ambiguity is also a remote heir of the picture theory. Trees are schemata and schemata are stylized pictures.

Trees play a role in semantics insofar as they account for syntactic *ambiguity*. The difference in meaning between the two readings of the phrase 'beautiful girl's dress', for instance can be easily captured by means of two trees

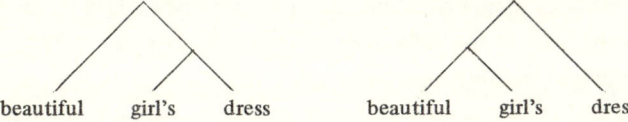

There are, however, more subtle ambiguities which cannot be dissipated by *segmentation*. In his *Introduction to Theoretical Linguistics* (1968), Lyons gives the following example

>   Amor Dei,
>   (Deus amat ....),
>   (.... amat Deum).

The ambiguity in this syntagm has traditionally been given an explanation in which

the phrase *amor Dei* is related to, and indeed in some sense derivable from, two sentences: (i) a sentence in which *Deus* ... is the subject of the verb *amare* ...; (ii) a sentence in which *Deum* ... is the object of the verb *amare* ... .[29]

In other words, to analyze a compound like 'Amor Dei' we need something *more* than a tree, we need *transformations* (in the sense in which this word is used in transformational grammar) which enable us to pass from one tree to another. As Quine puts it in *Philosophy of Logic* (1970), "Some compounds are best analyzed by working back and forth between different trees of construction, and transformations provide for this lateral movement".[30]

If we regard trees as a remote heir of the picture theory, we can even move a step further: transformations which relate trees and enable us to connect the surface structure tree with the deep structure tree are still in line with the picture theory insofar as much emphasis is put on trees. But sooner or later we will reach a point where the picture theory cannot be amended or extended any more but has to be simply discarded.

The need for an alternative theory already begins to be felt when grammatical relations are taken into consideration. If we regard *trees* as a remote heir of the picture theory, we might be tempted to argue that *transformations* are also related to that 'paradigm' (in the Kuhnian sense) insofar as they are defined in terms of trees. A closer look at the matter, however, reveals that such an account is an oversimplification. Let us see why this is so.

Transformations are semantically important precisely because they enable us to retrieve deep structure in which *grammatical relations* show themselves in a transparent manner. In the example supplied by Lyons, the word corresponding to 'dei' in the deep structure bears the *subject relation* to the verb corresponding to 'amare' in the former deep-structure tree and the object relation in the latter.

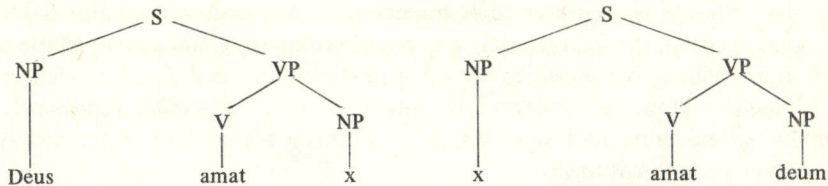

But now the question arises of explaining how *grammatical relations*, as opposed to syntactic factors such as word order contribute to the meaning of the sentences? Can the picture theory offer a solution to this problem? As a preliminary answer we will consider the Katz and Fodor solution.

In 'The Structure of a Semantic Theory' (1963),[31] J. Katz and J. Fodor introduced into Semantic Theory a new component which was supposed to explain the contribution of grammatical relations to meaning, namely a set of *projection rules*. As Katz wrote later:

> A projection rule applies to a set of readings for constituents that are grammatically related and there is a different projection rule for each distinct grammatical relation. So there is one projection rule for attribution, one for the subject-predicate relation, one for the verb-object relation, and so on.[32]

Projection is conceived as an operation of amalgamation of *trees* and this amalgamation is sometimes described in obscure metaphoric terms. Consider, for instance, this statement made by Katz: "the projection rule that combines for a verb and its object forms the derived reading by embedding the reading of the latter into that of the former at a fixed position".[33] What does 'embedding' mean when applied to readings, i.e. to meanings?

As a preliminary step towards answering that question, let us consider what projection does. If we look at Katz and Fodor's own examples, what it does is reduce ambiguity by cancelling senses the combination of which would create nonsense. Take for instance the lexical item 'ball'. It comprises the three readings or senses:

(1) ball = social activity for the purpose of social dancing;
(2) ball = physical object having globular shape;
(3) ball = solid missile for projection by engine of war.

If you combine 'ball' with 'hits', the first reading is cancelled out. *Projection* is thus no more than *selection*.

Langendoen[34] explicitly acknowledges this equation in *On Selection, Projection, Meaning and Semantic Component* (1967), where he says that grammatical selection and semantic projection are one and the same thing.

The trouble is, however, that projection understood as selection is too vague a notion to characterize the contribution of grammatical relations to the meaning of sentences which varies with the grammatical relations considered. Moreover selection is purely negative: it *excludes* unwanted combinations but does not *provide* a positive account for semantically acceptable combinations.

We are forced to the conclusion that Katz and Fodor's semantics does not offer a satisfactory account of the role played by grammatical relations in the production of meanings. If we were left with their account, we would be justified in sharing Lyon's scepticism expressed in *Introduction to Theoretical Linguistic* (1978):

...we are at present unable to interpret the term 'product' (or 'compositional function' – to employ the more technical term) in the proposed definition of the meaning of a sentence or phrase as 'the product of the senses of its constituent lexical items'.[35]

There is, however, an alternative to Katz, Fodor and Langendoen's approach but this alternative takes us far away from the picture theory. This unquestionably new approach is worth examining because it raises ontological problems connected with the Realism-Nominalism controversy.

9. THE RECURSIVE DEFINITION OF TRUTH AS A TOOL FOR COMPOSITIONAL SEMANTICS

Davidson fully recognizes the need for a recursive semantics parallel to, but distinct from, syntax: "The hopeful thought is that syntax... will yield semantics when a dictionary giving the meaning of each syntactic atom is added".[36] But these hopes prove deceptive: "recursive syntax with dictionary added is not necessarily recursive semantics".[37] Davidson thinks that a recursive definition of truth à la Tarski will offer him the recursive and compositional semantics he seeks: "a theory of meaning for a language L show's 'how the meanings of sentences depend upon the meanings of words' if it contains a (recursive) definition of truth-in-L".[38]

How can a recursive definition of *truth* also offer a recursive and compositional definition of *meaning*? This can be explained in this way. In his *Introduction to Logical Theory* (1952), Strawson says that "to know the meaning of a sentence of the kind of statement-making sentence is to know under what conditions someone who used it would be making a true statement.".[39] In other words, to know the *meaning* of a statement-sentence is to know the *truth-conditions* of the *sentence*. Once we equate sentential meaning with truth-conditions, we are bound to admit that a recursive definition of truth which would represent how the truth-conditions of a complex sentence depend on the truth-conditions of the elementary sentences would, *ipso facto*, represent how the meaning of a complex sentence depends on the meanings of its parts when the latter are also sentences. This is precisely what a recursive definition of truth does insofar as it contains a separate clause for each connective which says how this connective contributes to the truth value of the whole once the truth-value of the components is known.

Admittedly, we want more. We also want to be able to know the meaning of an atomic sentence once we know the meaning of its parts where the parts are not sentences and thus are not truth-bearers, but are predicates and individual variables which cannot be said to have *truth*-value or *truth*-conditions.

At this stage, one might fear that a recursive definition of *truth* be incapable to provide a recursive theory of meaning, i.e. a recursive semantics. Tarski, however, solved the problem by replacing the semantic predicate 'true' by another one, the predicate 'satisfies' which can be predicated both of closed sentences and of open-sentences, i.e. of predicates.

This major step has been well described by Barbara Partee. Consider, she writes, the sentence 'Every man loves a woman such that she loves him', there is no way to determine truth-conditions for the embedded clause 'she loves him' in isolation. This raises a difficult problem. "Tarski's solution was to define the syntactic class of *formulas* to include both (closed) sentences and expressions like sentences but including true variables, and to recursively define the notion of *satisfaction* for formulas".[40] She continues: "The satisfaction conditions for the embedded formula x_1 loves x_2 then can be shown to contribute systematically to the truth-conditions for the complete sentence".[41] And she concludes "The power of this approach lies in the fact that for any language of first-order predicate logic it is possible (1) to give a finite recursive syntactic characterization of the set of all *formulas* of the language, (2) to give a finite recursive semantic characterization of *satisfaction* for all the formulas, based on that syntax, and (3) to define *truth* in terms of *satisfaction* in a way that leads to the correct characterization of truth conditions for the *sentences* of the language, i.e. these formulas that contain no free variables".[42]

10. RECURSIVE SEMANTICS AND NOMINALISM

Davidson wants a semantics which matches syntax as far as *rules* are concerned — the clause of the recursive definition of satisfaction correspond to the clauses of the recursive definition of well-formed formula — but not as far as *categories* are concerned. Davidson rejects a semantics which would posit classes of independent denotata for classes of terms of all syntactic categories.

He subscribes to the 'Fido'-Fido theory *only* for the individual constants:

> We decided a while back not to assume that parts of sentences have meanings except in the ontologically neutral sense of making a systematic contribution to the meaning of the sentences in which they occur. Since postulating meanings has netted nothing, let us return to that insight. One direction in which it points is a certain holistic view of meaning. If sentences depend for their meaning on their structure, and we understand the meaning of each item in the structure only as an abstraction from the totality of sentences in which it features, then we can give the meaning of any sentence (or word) only by giving the meaning of every sentence (and word) in the language. Frege said that only in the context of a sentence does a word have meaning; in the same vein he might have added that only in the context of the language does a sentence (and therefore a word) have meaning.[43]

Davidson's argument can be spelled out in this way: The smaller place you give to autonomous meaning, the larger place you give to heteronomous meaning. If predicates do not have reference (denotation), they have to be given a syncategorematic status which frees the use of such expressions from ontological commitment. If an expression has not meaning in itself, it can only contribute to the meaning for a larger linguistic entity. Which entity? Shall we take the sentences to be the units of meanings? This answer Davidson rejects and rightly so: it would not take into account the fact that once we fully know the meaning of a sentence (let say $A > B$) we also know the meaning of all sentences built with the same words ($B > A$). This is, I suggest, the reason why he takes language as a whole to be the unit of meaning.

In a later work 'In Defense of Convention T' (1973), he nevertheless imposes limits to his holism:

> Since there is an infinity of T-sentences to be accounted for, the theory must work by selecting a finite number of truth-relevant expressions and a finite number of truth-reflecting constructions from which all sentences are composed. The theory then gives outright the semantic properties of certain of the basic expressions, and tells how the constructions affect the semantic properties of the expressions on which they operate.[44]

But the ontological commitment to meanings or references is reduced to a minimum as shown by the end of the quotation which reads as follows:

> In the process, some expressions are taken as performing referentially — the variables, at least, but no expressions need to be regarded as naming or denoting anything unless they are unreduced singular terms.[45]

Davidson's recursive semantics perfectly fits in with a nominalistic theory of proposition. No intensional meanings are posited as entities and extensional meanings — references or denotations — are ascribed to individual constants only. To put it in a nutshell, Davidson contents himself with the minimal ontology which is involved in the semantics of a first-order predicate calculus.

Unfortunately, a semantics for predicate calculus cannot be extended to cover reasonably large fragments of natural language. Its fundamental weakness lies precisely in its inability to account for *grammatical relations*. Predicate calculus duly represents the difference between transitive verbs and intransitive verbs — the former are 2-place predicates, the latter are 1-place predicates — but it cannot represent the distinction between subject and object. This limitation was already pointed out by Church in 'Ontological Commitment' (1958):

> For example, common sense holds that a man who does not love at one time and does love at a later time has become or changed; but if somebody at one time does not love him and later does love him, he has not therefore necessarily changed... But there is not presently available a sound and adequate logic which maintains this ordinary language distinction between the subject and the object of a verb.[46]

The logic whose absence Church deplores has become available both in the works of the later R. Montague and elsewhere. In *Logical Types for Natural Language* (forthcoming: Reidel) Keenan and Faltz have spelled out a syntax and a semantics for natural language which is alive to the difference between grammatical relations.

In the above-mentioned monograph, predicate calculus is abandoned in favor of categorial grammar which recognizes categories traditionally used in the grammar of natural language such as verbs, adverbs, prepositions. The correlation between syntax and semantics is maintained. Expressions of each syntactic category are correlated with denotations belonging to a semantic category (logical types). For instance, intransitive verbs are correlated with a certain kind of functions, i.e. with homomorphisms which take their arguments in the set of individuals (conceived as sets of properties) and their images in the propositional algebra. Transitive verbs are correlated with homomorphisms which take their arguments in the set of individuals and their values in the algebra of intransitive verbs.

Syntactically speaking, transitive verbs are functions which take determined terms as argument and yield intransitive verbs. Verbs are function which take determined terms as argument and yield sentences. Hence *grammatical objects* can be distinguished from *subjects*: the former are arguments of transitive verbs, the latter arguments of intransitive verbs. The syntactic operation of combining an expression of one category with an expression of another is generally conceived as a functional application of members of one syntactic category on members of another. Functional application in syntax is matched by functional application in semantics. The latter accounts for the semantic combination of the denotation of one expression with that of another.

11. CATEGORIAL GRAMMAR, SET THEORETIC SEMANTICS AND NOMINALISM

Keenan's and Faltz' account is more flexible than Davidson's account but it requires much more powerful ontology: contrary to what is the case in Davidson's semantics, all categorematic expression: verbs, adverbs, prepositions, sentences receive distinct *denotations* which are sophisticated set-theoretic constructs (most of the time, functions of some sort). Can a semantics built along these lines still be counted as nominalistic?

Some concessions to an extensionalistic Platonism have undoubtedly to be made at this stage. But the following considerations can be made to mitigate our commitments: the model built up to account for the meaning of linguistic expressions should not be seen as a schematic representation of the *real world* but as a representation of *meaning*. For instance, to know the meaning of 'John walks' one has to know a semantic rule which computes the denotation of this sentence for any arbitrary model whatever, i.e. for a model in which the denotation of 'walks' contains the denotation of 'John' and also for a model in which the opposite situation obtains. Which of the two models is actualised is not a task for the semanticist to say. To that extent, denotations are not realities on their own which would stand out there along with stars, dogs, 'cabbages and kings'. Nevertheless our semantical metalanguage has to quantify over those set theoretic objects, and to that extent, we are committed to assume them. This breach in the nominalistic program has to be recognized.

12. GAME-THEORETICAL SEMANTICS

Keenan's and Faltz' semantics, or, for that matter, Montague's or Hausser's are based on Frege's compositionality principle.[47] Hintikka, however, has raised important objections against that principle understood as "the principle which says that the meaning of a complex expression is a function of its constituent parts". Among other things, he showed that this principle is supported by rather poor arguments. For instance, those who try to motivate the compositionality principle have to assume that meaning is context independent to a large extent.

Now, the assumption of context-independence, Hintikka observes, is closely related with another thesis (which) says that the proper direction of semantical analysis is from inside out in a sentence or other complex expression and this other thesis is undermined by counter-examples in natural language. As Hintikka showed in 'On the Proper Treatment of Quantifiers in Montague Semantics' (1974), "the way in which a semantical object is correlated with 'any x' must depend on the context in which this expression occurs".[48] Hence an 'outside-in' principle applies here.

This led Hintikka to propose an alternative approach: the game-theoretical semantics which retains some characters of model theoretic semantics: "In my game-theoretical semantics the meaning of a complex expression E is typically analyzed in terms of the meanings of certain simple expressions $E_1, E_2 ... E_j$". But he adds "even though they (the meanings of $E_1, E_2 ... E_j$) are obtained from E through certain game rules and are simpler than E, they are not always themselves *parts*".[49]

Hintikka's way of tackling the problem formulated by Lyons seems more adequate than Montague's and possesses the additional advantage of avoiding the reification of meanings understood as *parts* which is a conspicuous feature on Montague's solution. But continuing with this would take us too far afield.

REFERENCES

[1] A. J. Ayer, *Thinking and Meaning*, London, Lewis, 1947, p. 14.
[2] A. J. Ayer, 'Meaning and Intentionality', *Atti del XII Congresso Internazionale di Filosofia*, Relazioni introduttive, Sansoni, Firenze, 1958, p. 147.
[3] Husserl, *Logical Investigations*, transl. by J. N. Findlay from the second edition, Vol. 1, p. 330.
[4] M. Combes, *Le Concept de concept formel*, Publ. de la Fac. des Lettres et Sciences humaines de Toulouse, Série A, tome B, 1969, p. 11.
[5] Combes, *Ibid.*, p. 12.
[6] Combes, *Ibid.*, p. 12.
[7] L. J. Cohen, *The Diversity of Meaning*, Methuen, London, p. 13.
[8] A. J. Ayer, *Meaning and Intentionality*, cf. n. 2, p. 151.
[9] Ayer, *Ibid.*, p. 152.
[10] R. Wells, 'Meaning and Use', *Word* X (1954) 236.
[11] A. Church, 'Logic and Analysis', *Atti del X Congresso Internazionale di Filosofia*, Sansoni, Firenze, 1961, Vol. V, p. 77–81.

CHAPTER VII

[12] G. Ryle, 'The Theory of Meaning', Ch. E. Caton (ed.), *Philosophy and Ordinary Language*, p. 151–152.
[13] Ryle, *Ibid.*, p. 151.
[14] J. Vuillemin, *Leçons sur la première philosophie de Russell*, Colin, Paris, 1968, p. 80.
[15] Ryle, *Op. cit.*, p. 142.
[16] G. Ryle, 'Use, Usage and Meaning', *Proceedings of the Aristotelian Society*, Supplementary Volume, 35, (1961) p. 228.
[17] Ryle, *Ibid.*, p. 224.
[18] E. Benveniste, 'Les Niveaux de l'analyse linguistique', *Proceedings of the 9th International Congress of Linguistics*, Mass., 1962, Mouton and Co., 1964.
[19] G. Frege, *Die Grundlagen der Arithmetik*, Breslau, 1889, trans. J. L. Austin, *Foundations of Arithmetic*, Blackwell, Oxford, 1952, p. 73.
[20] L. Wittgenstein, *Notebooks* (1914–1916), Blackwell, Oxford, 1961, p. 93, quoted by P. Wienpahl in *Wittgenstein and the Naming Relation, Inquiry*, No. 4, 1964, p. 341.
[21] R. Carnap, *Meaning and Necessity*, Chicago University Press, 1947, p. 30.
[22] G. Ryle, 'Discussion', *Philosophy* (1949) 75.
[23] R. Gale, 'Propositions, Judgments, Sentences and Statements', *Encyclopedia of Philosophy*, Colliers MacMillan, London, 1967, VI, p. 502.
[24] Wittgenstein, *Op. cit.*, p. 41.
[25] Wittgenstein, *Op. cit.*, p. 45 ff.
[26] J. C. Morrison, *Meaning and Truth in Wittgenstein's Tractatus*, Mouton, The Hague, 1968, p. 61.
[27] J. F. Rosenberg, 'Wittgenstein's Theory of Language as Picture', *American Philosophical Quarterly* 5, (1968) 25.
[28] F. Vandamme, *Simulation of Natural Language*, Mouton, The Hague, 1972, p. 60.
[29] J. Lyons, *Introduction to Theoretical Linguistics*, Cambridge, University Press, 1968, p. 249.
[30] W. V. Quine, *Philosophy of Logic*, Prentice-Hall, 1970, p. 17.
[31] J. Katz and J. Fodor, 'The Structure of a Semantic Theory' *Language* 39, (1963) 170–210.
[32] J. Katz, 'Recent Issues in Semantic Theory', *Foundations of Language* 3 (1967) 128.
[33] Katz, *Ibid.*, p. 128.
[34] T. Langendoen, 'On Selection, Projection, Meaning and Semantic Content', *Working Papers in Linguistics*, Mimeo, The Ohio State University, 1967, p. 102.
[35] Lyons, *Op. cit.*, p. 476.
[36] D. Davidson, 'Truth and Meaning', *Synthese,* 17, (1967), repr. in J. W. Davis, D. J. Hockney and W. K. Wilson (eds.), *Philosophical Logic,* D. Reidel, Dordrecht, 1969, p. 4.
[37] Davidson, *Ibid.*, p. 5.
[38] Davidson, *Ibid.*, p. 7.
[39] P. F. Strawson, *Introduction to Logical Theory*, Methuen, London, 1952, p. 211.
[40] B. Partee, 'Montague Grammar and Transformational Grammar', *Linguistic Inquiry* 6, (1975), p. 208.
[41] Partee, *Ibid.*, p. 208.
[42] Partee, *Ibid.*, p. 208.
[43] Davidson, *Op. cit.*, p. 5.

[44] D. Davidson, 'In Defense of Convention T' in H. Leblanc (ed.), *Truth, Syntax and Modality*, North-Holland, 1973, p. 81.
[45] D. Davidson, *Ibid.*, p. 81.
[46] A. Church, 'Ontological Commitment', *Journal of Philosophy* 55, (1958) 1011.
[47] See R. Montague (ed.), *Formal Philosophy*, and with an Introduction by R. Thomason, Yale University Press, 1974, p. 159.
[48] J. Hintikka, Theories of Truth and Learnable Languages. (Unpublished.)
[49] J. Hintikka, 'On the Proper Treatment of Quantifiers in Montague Semantics' in S. Stenlund (ed.), *Logical Theory and Semantic Analysis*, Reidel, Dordrecht 1974, p. 56.

CHAPTER VIII

AN ATTEMPT AT A NEW SOLUTION FOR THE ENIGMA OF THE MEANING OF FALSE SENTENCES

1. CONDITIONS OF ADEQUACY ON A SATISFACTORY ANSWER

At first glance, the question 'What is the meaning of false sentences?' looks like a *pseudo-question*, much like the parallel question 'What is the meaning of true sentences?', against which Ayer objected that it presupposes that all sentences mean the same. But it is possible to interpret these questions differently. The person who asks them does not necessarily seeks a *single correlate* as the meaning of all false sentences. He may rather be preoccupied with the question to which *category of reality* the correlates of true and – if they exist – false sentences belong. Formulated in this way, my question is not founded on absurd presuppositions.

The difficulty of the enigma I am trying to solve lies in the fact that one is caught between two contradicting exigencies: on the one hand, one must account for the *invariance* and the *neutrality* or the 'symmetric and median position' of the *meaning* of sentences with respect to their *truth-value*, that is, the fact that truth-value does not determine meaning; on the other hand, one must account for the *asymmetry* of the false in relation to the true, that is, for the fact that the false is parasitic in relation to the true, whereas the *negative* is not parasitic in relation to the *affirmative*.

There is no doubt about this asymmetry of the false in relation to the true: the false sentence must be considered a *bad* description of reality, rather than a *good* description of the *unreal*. The false differs from *fiction* precisely in that it *claims* to be true. Now, to admit this truth in the proper sense is to recognize the *asymmetry* in question, for a bad description can be judged as such *only* in relation to a norm, and an *infringement* is logically posterior to the norm in relation to which it is defined.

The theories presented thus far have satisfied either the first or the second exigency, excluding one or the other, whereas they should have satisfied both simultaneously.

The theory of the proposition advanced by Russell in *Inquiry into Meaning and Truth* (1940), satisfies the first condition of adequacy to the exclusion of the second. It accounts for the invariance of the meaning of a sentence to its truth or falsity by sliding, like an interpolated sheet of paper, between a sentence and reality a *neutral entity* of a psychological sort: the proposition: "... sentences signify something other than themselves, which can be the same when the sentences differ. That this something must be psychological (or physiological) is made evident by the fact that propositions can be false".[1]

Russell's solution does not explain how false propositions *infringe* on linguistic conventions of a certain kind.

In 'Are there Propositions' (1929–1930), Ryle presents a theory of meaning which, on the contrary, satisfies the second condition of adequacy to the exclusion of the first. Although this is an old theory which Ryle abandoned, it still deserves a thorough examination because it reveals certain aspects of the problem better than other less vulnerable theories.

2. RYLE'S SOLUTION TO THE ENIGMA OF THE MEANING OF FALSE SENTENCES

Ryle approaches the problem of the meaning of sentences indirectly by way of an analysis of the intellectual operations performed by a listener. He enquires about the nature of the *understanding* by the mind of a statement.

For Ryle, the understanding of the sense of a sentence is a form of knowledge, linguistic knowledge:

...understanding the statement 'X is Y' is a case of knowing; not knowing that X really is Y, but knowing about the statement 'X is Y' that it is as if (it has the character that it would have if) X is (or were) Y.[2]

Ryle means with this sentence that to understand the statement 'X is Y' is to know a fact of the following hypothetical form: statement 'X is Y' would be true if X were Y.

What is the *reference* of sentences according to Ryle? It is neither a *proposition*, as it is in Carnap's *Introduction to Semantics* (1942), nor a truth-value, as argued by Frege and Church. The reference of a sentence is the *fact* to which it corresponds if it is true.

But how in this case does one explain the meaning of *false* sentences since here there is no fact for them to state? This objection was forcefully stated by A. J. Ayer in *Thinking and Meaning* (1947):

Thus, if I believe truly that it is going to rain to-morrow, the fact that it will rain to-morrow ... may ... be said to be what the words 'it is going to rain to-morrow' mean. But suppose that my belief is false, as it very well may be. In that case, the fact that it will rain to-morrow cannot be the object of my belief, for the very good reason that there is no such fact. And for the same reason, it cannot be what the words 'it is going to rain to-morrow' mean.[3]

Ryle resorts here to an heroic solution: he denies that false sentences have meaning. He writes: "In the strict sense, only those statements *mean* something, which state a fact (to someone who knows the fact)".[4] Further, he continues: "So a statement which does not state a fact known to us does not genuinely symbolize a reality and is therefore only a quasi-symbol".[5] This solution does not fulfil one of the two adequacy conditions. It should therefore be rejected.

3. THE POSSIBILITY OF FALSITY AS A BY-PRODUCT OF THE CREATIVITY OF LANGUAGE

The founder of the Vienna Circle, M. Schlick, very clearly formulated the idea that human language is creative, and further the idea that this property is explained by the contribution of syntactic factors to meaning. He writes:

> The essential characteristic of language, ... is its capability of expressing facts, and this involves the capability of expressing *new* facts, or indeed *any* facts ... . The same set of signs which was used to describe a certain state of affairs can, by means of rearrangement, be used to describe an entirely different state of affairs *in such a way that we know the meaning of the new combination without having it explained to us.*[6]

Schlick, in offering an explanation of the immediate intelligibility of sentences we have never heard before, provides us at the same time with part of the solution of the enigma of the meaningfulness of false sentences.

In the preceding chapter, we saw that syntactic factors, like the *scope* of parentheses, have the strange property of contributing to reference *without themselves having reference*. With the intervention of syntactic factors, the narrow bond linking sign to signified object *loosens*, the connection becomes more *indirect*. Whereas in the case of non-fictional proper names meaning is somehow riveted to the present or past existence of the denoted object, and depends on the prior existence of the referent — a point which the causal theory of proper names has emphasized — in the case of sentences the direction is reversed, the proposition or possible state of affairs expressed by the sentence is generated or, at least, rendered conceivable by the sentence, i.e. sense *is projected from signs to things rather than from things to signs*.

From this one may explain that things may not answer to the signs, and that signs may function vacuously. The possibility of forming sentences which carry meaning, but which are false, and the possibility of forming original sentences that are nevertheless immediately intelligible, are almost like the two sides of a coin. Not exactly however, since, as F. Vandamme[7] observed, we could have a very poor language in which *no* original sentences are expressible, although false ones are. By bringing this coincidence to light we believe that we have solved in part the enigma of false sentences. We have, in effect, explained how it is *possible* that sentences be at the same time meaningful and false.

This is, however, only part of the solution. To be complete, one would have to explain how this possibility is *realized* despite the fact that the speaking subject normally aims at truth. In Section 5 we shall offer our solution to this problem. This solution will no longer require a recourse to *syntax*, but rather to the *pragmatic* dimension of meaning.

4. THE SOLUTION OFFERED BY POSSIBLE WORLDS SEMANTICS TO THE ENIGMA OF THE MEANING OF FALSE SENTENCES

In 'Semantic competence' (1978), Cresswell presents an account of the meaning of sentences which avoids the objections raised against the previous study.

In possible-worlds semantics a collection of possible worlds is taken as primitive and the meaning of every sentence is identified with the set of worlds in which it is true. A set of possible worlds is therefore sometimes called a *proposition*.[5]

This account neatly distinguishes between *truth-value* and *meaning*, the latter being equated with *truth-conditions*. Moreover it entails consequences which are consonant with our intuition. For instance, it entails as we shall see that knowing the *truth-value* is neither a *necessary* nor a *sufficient* condition for knowing the truth-conditions, that is to say the meaning. And this is as it should be.

Knowing the truth-conditions amounts to being able to divide the set of possible worlds into two subsets: the set in which the sentence is true and the set in which it is false. But this does not equip us with the capacity of telling which of the possible worlds is the *actual* world. On the other hand, knowing the truth-conditions involves a consideration of possible worlds which is not required for knowing the truth value.

Cresswell's account of meaning in terms of truth-conditions might seem to conflict with a nominalistic outlook in so far as it calls upon possible worlds, even if it treats the latter as primitive theoretical concepts. Moreover, Cresswell's account seems to clash with what genetic psychology suggests, i.e. with the idea that the notion of possibility is generated by operations with signs and not the other way around. Before examining Cresswell's account on its own grounds, let us consent to a short digression in order to show how it can be reinterpreted so as to avoid these objections.

It is reasonable to suppose that the first sentences we learn are, to use Quine's terms, the occasional observational sentences, i.e. those sentences which describe a short event observable by the speaker and hearer and more or less contemporary with the utterance of the sentence (cf. the well known example: 'Lo, a rabbit!'). When confronted with the task of accounting for the meaning of these sentences we can take up Cresswell's definition and substitute 'circumstances in which to count is true' for 'possible worlds in which the sentence S is true'.

As to the acquisition of the meaning of the so-called eternal sentences such as 'snow is white', it can be explained in term of operations on signs as Quine suggests in *Roots of Reference* (1974):

> In learning an occasion sentence we learn in what circumstances to count it true and in what circumstances false. In learning the eternal predicational construction, we are learning how to judge whether a given pair of terms produces a true predication, true for good, or a false one, false for good.[9]

The difference is that "the variability of truth-value has withdrawn merely to a higher level of abstraction".[10] In other words, we learn the truth-conditions of an *occasion sentence* such as 'there, a rabbit is approaching' by varying the *circumstances*, uttering the sentence and querying assent. We learn the truth conditions of an *eternal sentence* by varying the *terms* in a standing sentence, uttering it and querying assent. In each case "we are learning how to distribute truth-value"[11] but we never have to travel across *possible worlds*.

Even if the truth-conditional account of meaning can be reformulated in this nominalist way however, it does not fulfill all our requirements: it does not comply with the second of them since it does not explain why truth is the *designated* or *preferred* truth-value.

5. A PRAGMATIC SOLUTION OF THE ENIGMA

A truth-conditional account of the meaning of declarative sentences is not false but incomplete. The semantic component of meaning is taken care of but the pragmatic component is left out of the picture. But, it is the pragmatic component which, as I shall try to show, explains why for a sentence to be *false* is a defect, while for it to be *negative* is not.

As a starting point, let us consider the following two linguistic games described in Waissmann's[12] posthumous work *The Principles of Linguistic Philosophy* (1965):

(a) "In the room where A sits there is a lamp which lights up red or

THE ENIGMA OF THE MEANING OF FALSE SENTENCES 129

green at irregular intervals. *A* is to watch the lamp and say which colour he sees". We have here an exercise in the application of colour names. The subject who errs here can only make a linguistic error.

(b) "This antithesis could occur in a game in which *A* is to *guess* which colour will come next". Here the subject can make either a factual error or a linguistic error.

Declarative sentences are used to convey guesses and conjectures just as well as contemplated 'facts', yet guesses as descriptions are assessed along the same dimension as descriptions, namely the True-False dimension. The sincerity condition mentioned in Chapter III (Cf. Searle's *Taxonymy of Illocutionary Acts*) comes next into play.

When we make a guess, we ideally should say both what is true and what we believe. But it may turn out that our beliefs are false. Since, ex hypothesi, we don't know that this is so, we have to live by our beliefs and speak in accordance with them. The institution of language is set up in that way.

If someone erroneously believes that all swans are white and utters the sentence 'all swans are white', he *misuses* at least one word to the extent to which he does not use it in accordance with the paradigm use. In the paradigm use, the sentence is both believed and true.

When we claim that false sentences embody a *misuse*, we do not say that they are *pseudo-symbols* as Ryle thought. Moreover we stress that such a misuse is somehow built into the language game of statements as can be gathered from the fact that the copula is liable to two sorts of mistake instead of only one as is the case for the other words in the sentence.

6. NOMINALISM AGAIN

For there being meaning anywhere there must be truth somewhere. For there being belief anywhere, there must be knowledge somewhere. These two views held by Price and Hamlyn, respectively, lie at the bottom of the previous account of the meaningfulness of false sentence.

Once again I wish to stress that sentences are related not to Platonistic entities but to the world. The relation however is highly indirect. Theoretical sentences for instance are related by a great number of tortuous connections to a few observational sentences whose meaning is taught ostensively in the celebrated 'paradigm cases'. Once again we see that it is not the *significata* of the sentences which are queer entities, it is the *relation* between the sentence and the world which is highly sophisticated.

CHAPTER VIII

REFERENCES

[1] B. Russell, *An Inquiry into Meaning and Truth*, Allen and Unwin, London, 1940, reprinted 1966, p. 189.
[2] G. Ryle, 'Are there Propositions?', *Proceedings of the Aristotelian Society*, XXX (1929–1930) 122.
[3] A. J. Ayer, *Thinking and Meaning*, H. K. Lewis, London, p. 3.
[4] Ryle, *Op. cit.*, p. 120.
[5] Ryle, *Ibid.*, p. 121.
[6] M. Schlick, *Gesammelte Aufsätze* (1921–1936) p. 153, quoted by F. Waissmann, *The Principles of Linguistic Philosophy*, Harré (ed.), MacMillan, London, 1965, p. 305.
[7] F. Vandamme, 'Le nominalisme et la proposition', *Communication and Cognition* (1977) 88.
[8] M. J. Cresswell, 'Semantic Competence', in F. Guenthner and M. Guenthner-Reutter (eds.), *Meaning and Translation*, Duckworth, 1978, p. 12.
[9] W. V. O. Quine, '*Roots of Reference*', 1973, Open Court, La Salle, Illinois. p. 265.
[10] Quine, *Ibid.*, p. 265.
[11] Quine, *Ibid.*, p. 265.
[12] F. Waissmann, *The Principles of Linguistic Philosophy*, Harré (ed.), MacMillan, 1965, p. 286.

CHAPTER IX

THE IDENTIFICATION CRITERION FOR PROPOSITIONS

1. THE IMPORTANCE OF FINDING A CRITERION OF PROPOSITIONAL IDENTITY

Now that we have surveyed the philosophical disciplines where the concept of the proposition plays a role, we must answer a critical question which our method prevented us from asking: what is the criterion of identity, the principle of individuation, for propositions?

The answer to this question is essential if we want the concept of the proposition to have a sound epistemological status. In general, definitions which do not satisfy such a criterion are not *operational*. As Quine remarks in *Word and Object* (1960), "A large part of learning 'apple' or 'river' was learning what counts as the same apple or river re-exposed and what counts as another".[1]

What is true of empirical concepts is also true for theoretical concepts. Quine continues: "Similarly for 'proposition': little sense has been made of the term until we have before us some standard of when to speak of propositions as identical and when as distinct. Not being physical, a proposition cannot, like an apple or river, be exposed and re-exposed; but it admits of something analogous".[2]

The demand for a criterion of propositional identity seems to be a requirement whose justification poses no problem. It has, however, been questioned. In 'Is the Concept of Referential Opacity Really Necessary?' (1963), Prior considers the expression '*p* is the same proposition as *q*' as a connective and not as a predicate of the metalanguage. He writes:

I do not see either that there is any obligation to *define* 'propositional identity' in the above-suggested sense – it is simply what is involved whenever we speak, e.g. of two people 'believing the same thing', 'saying the same thing', 'fearing the same thing' ('thing' is not, of course, a generalized *name* here but a generalized *sentence*) – ... . But although we need not define this connective ['believes that'] we ought to be able to give some laws for it... .[3]

These words of Prior's, however, ought not to dissuade us from trying to define a criterion of propositional identity; for the fact that *in the context of a particular problem* we can treat '*p* has the same meaning as *q*' as a primitive

131

132   CHAPTER IX

term does not spare us the *general obligation* to provide a criterion of propositional identity, if we wish to use the word 'proposition'. Quine's methodological dictum, consequently, retains all of its force.

2. THE DEFINITION OF PROPOSITION IN TERMS OF SYNONYMY

In *Inquiry into Meaning and Truth* (1940), Russell formulated a definition of proposition that provides us with a criterion of propositional identity. Furthermore, he sees in it the only possibility for such a definition:

> Since 'having the same significance' is a relation which can certainly hold between two sentences – e.g. 'Brutus killed Caesar' and 'Caesar was killed by Brutus' – we can make sure of *some* meaning for the word 'proposition' by saying that, if we find no other meaning for it, it shall mean 'the class of all sentences having the same significance as a given sentence'.[4]

Russell's criterion of propositional identity rests, as one sees, on the notion of *synonymy*. It is certainly synonymy at the level of sentences and not of words. Nevertheless, the question whether there exist synonymous *words* clearly bears on the question whether there are synonymous *sentences*.

Let us first ask whether there are synonymous words at the base of the same language. In *Cours de linguistique générale* (1915), Saussure gives a negative answer to this question.

> In a language all words which express closely related ideas limit each other reciprocally: synonyms such as *redouter, craindre, avoir peur* have their proper value only by contrast; if *redouter* did not exist its content would go to the others.[5]

In 'Le binarisme, concept moteur de la linguistique' (1969), M. Leroy brings out very clearly the basic tenet of Saussure's theory:

> The elements of language acquire their value only in contrast with each other, in so far as they do not get confused with each other, thus it is not a truly positive quality that characterises them, but their contrasting qualities and their different values.[6]

Thus, linguists leave us no hope of finding synonymies in language. Do they, however, admit that words belonging to *different* languages might be synonymous? No! The absence of a semantic isomorphism between different languages is well-established, and is often illustrated by the example of the semantic field of colours. The same objective reality – the colour spectrum – is, in fact, divided differently in different languages. That portion of the spectrum which is described in English by the four words *green, blue, grey* and *brown* is described in Gaelic by the three terms *gwyrrd, glas, llwyd*.

Must we conclude from this that the concept of synonymy is an empty one and that every criterion of propositional identity founded on it is irreparably condemned? This conclusion seems premature. Certain accomodations are in fact possible.

In 'On Likeness of Meaning' (1949),[7] Goodman reached the same conclusion as the linguists, but by another route, and he formulated a criterion of synonymy which was so narrow that no pair of non-*equiform* expressions could satisfy it. Goodman proposed that we speak of a *resemblance in sense* instead of an *identity of sense*, and consequently that we speak of a "degree of synonymy". In 'On Some Differences about Meaning' (1952),[8] he further showed how we could define the threshold from which synonymy is a useful notion.

However, considering the problem with which we are concerned, the notion of *resemblance in sense* cannot play the role assigned by Russell to *synonymy*, and for an obvious technical reason: the definition of a proposition as the class of sentences synonymous with a given sentence, as Russell proposed, is a *definition by abstraction*. In order to construct a definition by abstraction, one needs an equivalence relation. Whereas *identity* of meaning satisfies this requirement, *resemblance* of meaning does not. Resemblance is symmetric and reflexive, but it is not transitive.

There is, however, another possible qualification on the concept of synonymy. Instead of conceiving of synonymy as a matter of *degree* as Goodman does, one could *relativize* it to the context, preserving at the same time its character of absolute identity. This is the solution proposed independently by Naess in 'Synonymity as Revealed by Intuition' (1954),[9] and by Pêcheux in *Analyse automatique du discours* (1969).[10]

Pêcheux has illustrated this with examples which will not unfortunately translate into English. His point, however, can be made with the pair of words 'meaning' and 'sense'. These words are interchangeable in the contexts: 'John knows the meaning of the word 'oxymoron'', "analytic' has a recognized philosophical sense'. However, they are not interchangeable in the contexts: 'His look was full of meaning', 'He approached the problem with sense'. Note that one could combine the two methods and *quantify* the degree of synonymy of two terms as a function of the number of contexts in which they are interchangeable preserving grammaticality.

Pêcheux's suggestion, however, does not solve our problem, but only displaces it. We can certainly see that by *relativizing synonymy to context* we can halt the *undermining* which threatened the previous solution; but we must still be precise in what we understand by *context*, whether it is the sentence or an entire discourse. This latter is indefinitely extendable due to the fact that sentences may be as long as the creativity of language allows them to be.

The answer to this question reveals a crucially important fact. If absolute synonymy cannot be defined at the level of sentences but only at the level of language, then Russell's definition in terms of the synonymy of sentences would be irreparably compromised. Carnap has tackled this fundamental problem at the level of sentences and we shall now examine his decisive contribution to its solution.

3. INTENSIONAL ISOMORPHISM

Our search for a criterion of propositional identity has thus far turned out to be unsuccessful. A criterion founded on interdeducibility ("mutual entailment" in the Anderson and Belnap sense) is inapplicable to natural language. Russell's criterion based on *synonymy* is also unacceptable, though for other reasons. Are there other possibilities?

In *Meaning and Necessity* (1947), Carnap formulated a third criterion; this attempt is based on the notion of *intensional isomorphism*. Though it is also open to objections, this criterion must be closely examined because it introduces the notion of an isomorphism, and this notion will turn out to be useful when we ourselves attempt to provide a criterion of propositional identity. Carnap writes:

> Let us consider, as an example, the expressions '2 + 5' and 'II sum V' in a language S containing numerical expressions and arithmetical functors. Let us suppose that we see from the semantical rules of S that both '+' and 'sum' are functors for the function Sum and hence are L-equivalent; and, further, that the numerical signs occurring have their ordinary meanings and hence '2' and 'II' are L-equivalent to one another, and likewise '5' and V'. Then we shall say that the two expressions are *intensionally isomorphic* or that they have *the same intensional structure*, because they not only are L-equivalent as a whole, both being L-equivalent to '7', but consist of three parts in such a way that corresponding parts are L-equivalent to one another and hence have the same intension.[11]

He adds that,

> It seems advisable to apply the concept of intensional isomorphism in a somewhat wider sense so that it also holds between expressions like '2 + 5' and 'sum (II, V)' Because the use in the second expression of a functor preceding the two argument signs instead of one standing between them ... may be regarded as an inessential syntactical device.

This criterion of propositional identity based on intensional isomorphism has undeniable merit. It partly solves the problem of substitution into belief contexts, which posed a serious threat to Russell's definition of a proposition as a class of synonymous sentences quoted on p. 132. But, unfortunately, this criterion is open to the same kind of objection which was raised against the notion of synonymy.

The criterion in question appeals, in effect, to the notion of L-equivalence which, despite appearances, contains the *same defects* as the notion of synonymy. In Carnap's terminology, 'L-equivalent expressions' does not mean 'expressions whose equivalence is demonstrable by logic', but 'expressions whose equivalence is demonstrable by logic *enriched by semantic rules* of the language'. To convice oneself of this one has only to recall that for Carnap, "The property Human is L-equivalent to the property Rational Animal".[12] Now, this extended notion of L-equivalence is as vague as that of *synonymy* or *analyticity*, except in the case of the logician who confines himself to an *artificial language* where semantic rules are defined *arbitrarily*. However, the interest in such a language as a source of insight into the structure of a natural language remains to be demonstrated.

Our criticism of Carnap's criterion concerns only the notion of intension. We are, therefore, right in asking ourselves if this criterion could not be amended. For example, what would happen if in the statement of this criterion one replaced the word 'intensional' by the word 'extensional'? One would avoid the objections aroused by the imprecise notions of L-equivalence, synonymy and intensional identity. Would the thus amended criterion still play the role intended for it? Would it not be too liberal to provide a criterion of identity for propositions? We must now consider this question.

We shall try to demonstrate that the notion of *extensional isomorphism* offers a promising starting point in our search for a criterion of propositional identity.

4. THE ROLE OF THE NOTION OF ISOMORPHISM IN DEFINING A CRITERION FOR THE IDENTITY OF PROPOSITIONS

Before considering the proposed criterion, we must understand exactly what it asserts and what is the *raison d'être* for its constituents. More precisely, we must ask ourselves about the role played by the notion of *isomorphism* in the statement of this criterion. Why, for example, can one not be satisfied with a criterion of propositional identity founded on the notion of *coextensionality*?

The answer to this fundamental question will require the use of some technical notions whose definitions must be recalled before we deal with the subject itself. Four formulas are involved which can be set up as rules of inference.

(a) The definition of identity and the principle of substitutivity

$$x = y \cdot = :(\Phi) : \Phi!x \cdot \supset \cdot \Phi!y \quad \text{Df}$$

This is Def. 13.01 from *Principia Mathematica*. Russell comments on it as follows: "This definition states that x and y are to be called identical when every predicative function satisfied by x is also satisfied by y".

The rule of inference corresponding to this definition is Leibniz's principle of substitutivity.

(b) The general principle of extensionality

$$(A)(B)(f)\,[A = B \cdot \supset \cdot f(A) \equiv f(B)]$$

where A and B range over objects, classes or relations in extension and when f is a suitable predicate-expression.

Contrary to the first principle, this one is acceptable only with certain limitations. As soon as one leaves the field of mathematics it is easy to produce counter-examples. Thus, the class of chimeras is coextensive with the class of centaurs, but the class of images of chimeras is not necessarily coextensive with the class of images of centaurs.

(c) The principle of truth-functionality

$$(p)(q)(f)\,[p \equiv q \cdot \supset \cdot f(p) \equiv f(q)]$$

This is a special case of the preceding principle. It affirms that all functions of propositions are truth-functions. This principle, which we call the principle of truth-functionality, is clearly not universally applicable. Not all connectives are truth-functional. 'And' is, but 'because' is not.

(d) The principle of intensionality

$$(p)(q)(f)\,[p \bowtie q \cdot \supset \cdot f(p) \bowtie f(q)]$$

This principle corresponds to the principle of extensionality for modal contexts. It is obviously different, since it requires that the propositions to be interchanged be strictly equivalent, not simply materially equivalent.

This modest technical baggage is all we need to establish that *coextensionality between sentences* cannot be taken as a criterion of propositional identity. Indeed, with only the definition of identity and the principle of truth-functionality one can easily demonstrate that such a criterion would entail an absurd consequence, namely:

$$p \equiv q \cdot \supset \cdot p = q$$

Russell, who demonstrates this point in *Principia* (2nd edn.), remarks: "There will thus be only two propositions, one true and one false. This was Frege's point of view, but it is one which cannot be easily accepted".[13]

One might think that it is not necessary to introduce the notion of *isomorphism* to remedy this situation, and that it is simply enough to reinforce the requirement of coextensionality and demand that it hold at a finer level, at the level of the (categorematic) constituents of the sentence rather than at the level of the sentence taken as a whole. But this remedy does not suffice because, if one only demands coextensionality at some level, then a peculiar process of 'logical phagocytosis' is immediately launched, a process whereby, one after another, all differences of sense between sentences are inexorably erased, allowing only differences of truth-value.

Church has described this process with utmost clarity in his *Introduction to Mathematical Logic* (1956):

Again the sentence 'Sir Walter Scott is the author of *Waverley*' must have the same denotation as the sentence 'Sir Walter Scott is the man who wrote twenty-nine Waverley Novels altogether', since the name 'the author of *Waverley*' is replaced by another name of the same person; the latter sentence, it is plausible to suppose, if it is not synonymous with 'The number, such that Sir Walter Scott is the man who wrote that many Waverley Novels altogether, is twenty-nine', is at least so nearly so as to ensure its having the same denotation; and from this last sentence in turn, replacing the complete subject by another name of the same number, we obtain, as still having the same denotation, the sentence 'The number of counties in Utah is twenty-nine'.

Now the two sentences, 'Sir Walter Scott is the author of *Waverley*' and 'The number of counties in Utah is twenty-nine', though they have the same denotation according to the preceding line of reasoning, seem actually to have very little in common. The most striking thing that they do have in common is that both are true. Elaboration of examples of this kind leads us quickly to the conclusion, as at least plausible, that all true sentences have the same denotation.[14]

In order to block Church's 'slippery slope' argument just quoted, we believe that it is necessary and sufficient to add to the requirement of *termwise coextensionality* the further condition of *isomorphic coextensionality*. Indeed the first two sentences in Church's example are synonymous and termwise coextensive, but they are not isomorphic. They are thus not interchangeable if we impose isomorphic coextensionality as a condition on such interchange of sentences. This is why we shall retain the requirement of isomorphism.

(1) 'Sir Walter Scott is the man who wrote twenty-nine Waverley Novels altogether' has the form:

$$\text{Scott} = (\imath x)\,[(\,\exists\,!\alpha)(x \text{ wrote } \alpha \cdot W\alpha \cdot \alpha \in 29)]$$

where '$W\alpha$' signifies "is a collection of novels published under the name 'by the author of Waverley' ".

(2) 'The number, such that Sir Walter Scott is the man who wrote that many Waverley Novels altogether, is twenty-nine' has the form:

$$29 = (\imath y) \,[(\exists !z)\,(\text{Scott} = z \cdot z \text{ wrote } \alpha \cdot W\alpha \cdot \alpha {\in} y)]\,.$$

5. PRELIMINARIES TO THE APPLICATION OF THE CRITERION OF EXTENSIONAL ISOMORPHISM

The adoption of the additional clause we propose, that is the isomorphism clause, is sufficient, in the context of Church's argument which we examined to prevent the collapse of *coextensionality* to *material equivalence* or, in other terms, to prevent the reduction of extensionality to truth-functionality. Is it sufficient in all cases? An argument recently advanced by Føllesdal and Davidson would seem to prove the contrary. If their conclusions are inevitable, our isomorphism clause has only a limited effectiveness. It will only have delayed the blotting out of distinctions of meaning which we must preserve. It is therefore necessary to examine immediately this argument.

In 'Causality and Extensionality' (1969), Miss Anscombe[15] attributes paternity for the argument in question to Quine who advanced it in 'Three Grades of Modal Involvement' (1953). Actually, the first formulation of this argument goes back to Church's 1944 review of Carnap's *Introduction to Semantics*.

In this review, Church clearly indicated the scope of his argument:

However, if a language ... contains an abstraction operator '(λx)' such that '(λx)(...)' means 'the class of all x such that ...', then – independently of the question whether the language is intensional or extensional – it is possible to prove that the designata of sentences of the language must be truth-values rather than propositions.[16]

In other words, *the abstraction operator annuls all differences other than differences of truth-value*, that is, it effaces *en bloc* the differences of sense (of propositions) with which Church is preoccupied, the differences between modal and extensional contexts in which Quine is interested, and finally the distinction between materially equivalent propositions and isomorphically coextensive propositions which we are trying to make out.

Fortunately, there is a way to block this threat to our criterion of propositional identity founded on extensional isomorphism. To find this way we should review carefully the Quine–Church reasoning in order to discover which assumptions are required to avoid its consequences.

THE IDENTIFICATION CRITERION FOR PROPOSITIONS 139

Consider a theory that at the same time contains the logic of classes as well as the abstraction operator, and in which the admitted principle of interchange is the *principle of intensionality*, that is the principle according to which two formulae are interchangeable in all contexts *salva veritate* if they are *logically* equivalent (if they *strictly* imply one another).

Church and Quine show that under these conditions the use of the *principle of the substitutivity of identicals* allows us to 'loosen the grip', to annul the restrictions imposed by the principle of intensionality in such a way that the theory tolerates a much more liberal interchangeability principle, that is, the principle of extensionality according to which two formulae are interchangeable *salva veritate* when they are *materially* equivalent.

These are the steps in the reasoning:

(1) Suppose that p and q have the same truth-value, that is, are *materially* equivalent:

$$p \equiv q.$$

(2) If p is true and does not contain a free occurrence of α, 'p' is *logically* equivalent to a statement affirming that the class determined by the propositional function

$$\alpha = \alpha \cdot p$$

is equal to the universal class. In formal terms:

$$p \leftrightarrow [\hat{\alpha}(\alpha = \alpha \cdot p) = V].$$

(3) If p is true, one obtains $\hat{\alpha}(\alpha = \alpha \cdot p) = V$. Let us make the same with 'q' true, one will obtain $\hat{\alpha}(\alpha = \alpha \cdot q) = V$. Since $\hat{\alpha}(\alpha = \alpha \cdot p)$ and $\hat{\alpha}(\alpha = \alpha \cdot q)$ are equal to V, one can infer

(4) $\quad \hat{\alpha}(\alpha = \alpha \cdot p) = \alpha(\alpha = \alpha \cdot q).$

(5) In the context $F(p)$, which is supposed true by hypothesis, we can replace 'p' by a formula which is logically equivalent to it. We thus obtain:

$$F[\hat{\alpha}(\alpha = \alpha \cdot p) = V].$$

(6) In virtue of the principle of the substitutivity of identicals and of the identity (4), we obtain:

$$F[\hat{\alpha}(\alpha = \alpha \cdot q) = V]$$

(7) In virtue of the principle of intensionality and of the following strict equivalence:

$$q \leftrightarrow [\hat{\alpha}(\alpha = \alpha \cdot q) = V]$$

we obtain from (6):

$F(q)$.

Everything proceeds as though we had admitted into our theory the principle of extensionality rather than that of intensionality.

How were the constraints imposed by the principle of intensionality avoided? The source of this circumvention is easy to identify. It is to be found in the principle of the substitutivity of identicals together with a fact propitiously pointed out by Quine in 'Three Grades of Modal Involvement' (1953), that is: "For classes, properly so-called, are one and the same if their members are the same — regardless of whether that sameness be a matter of logical proof or of historical accident".[17] Because for classes there is no distinction between necessary and contingent identity, classes play the role of a Trojan horse that destroys distinctions which one believed to be assured once and for all by making the interchangeability of formulae subordinate to the principle of intensionality.

The diagnosis of the trouble and the description of its causes indicate its remedy. In order to avoid the collapse of distinctions which one tries to enforce, either with the help of the principle of intensionality or with the help of the principle of extensional isomorphism, one may consider a *language which has been purified*, a language which does not contain an abstraction operator, a description operator or class names. Thanks to well-known classic procedures, it is easy to contextually eliminate descriptions, class abstraction and even class names, which one may consider along with Russell to be incomplete symbols. But there is an even easier way out which does not require jettisoning definite descriptions and class abstraction, namely to observe that 'p' is not extensionally isomorphic to '$\hat{\alpha}(\alpha = \alpha \cdot p) = V$' in

'$p \Leftrightarrow \hat{\alpha}(\alpha = \alpha \cdot p) = V$'.

In any case, in the canonical and austere language, whose construction materials are predicates and quantified variables, any interchange which is allowed by virtue of the principle of extensionality is also allowed by virtue of the principle of the substitutivity of identicals; but the converse is not true. This asymmetry is very important for us. Føllesdal calls our attention to it in 'Quine on Modality' (1968).

He writes: There are contexts that are at the same time "*referentially transparent* constructions on general terms or sentences" while these are "*extensionally opaque*...".[18] In other words, in such a language there are constructions in which co-referential singular terms are interchangeable in accord with the principle of the substitutivity of identicals but in which general terms (and sentences) that are simply coextensive (in opposition to those which are coextensive *and* isomorphic or to those which are co-intensive) are not interchangeable *salva veritate*.

The asymmetry stressed by Føllesdal is important to us not only because it assures us that extensional isomorphism does not collapse to extensionalism pure and simple, but also because it points up certain distinctions which we had already judged to be of use in other circumstances.

From the fact that in such a purified language the interchangeability of co-referential expressions does not entail that of coextensive expressions,

...it is immediately clear that a satisfactory semantics for the modalities must distinguish between expressions which refer (singular terms) and expressions which have extension (general terms and sentences, the extension of a sentence being its truth-value). This means that a Fregean semantics, according to which all expressions are treated on a par as referring expressions cannot be a satisfactory semantics for modal logic. Neither can a Carnapian semantics in which all expressions are treated indiscriminately as having extensions.[19]

There is a convergence worth noting between Føllesdal's conclusions and those we arrived at when examining the problems connected with the creativity of language.

The criterion of extensional isomorphism comes close to the criterion of identity of proposition formulated by Lemmon in *Statements and Propositions* (1966), which reads as follows:

Let $S(a)$ be a sentence containing the uniquely referring expression a, and $T(b)$ be a sentence containing the uniquely referring expression b. For any uniquely referring expression e let $rc(e)$ stand for the reference of e in context c. Then $S(a)$ in c_1 is used to make the same statement as $T(b)$ is used to make in c_2 if, and only if, $rc_1(a) = rc_2(b)$, and for any x, $S(x)$ if and only if, $T(x)$.[20]

In other words, for Lemmon, two atomic sentences make the same statement when their referential expressions are co-referential and when their predicate expressions are coextensive. Lemmon, therefore, has good reason for affirming that his criterion is "in textbook jargon, a purely extensional rather than an intensional notion".

Difficulties remain, however, when one tries to use the principle of extensional isomorphism to deal with belief sentences. There is nothing in that principle which prevents the sentences 'Of Cicero, John believes-true 'x denounced Catilina" and 'Of Tully, John believes-true 'x did not denounce Catilina" from ascribing inconsistent beliefs and yet the sentences 'John believes Cicero denounced Catilina' and 'John believes Tully did not denounce Catilina' do not ascribe inconsistent beliefs as opposed to ignorance in ancient history.

6. SOME FINAL REFINEMENTS OF THE NOTION OF EXTENSIONAL ISOMORPHISM

In *Problem fakt v logicke semantics* (1969), M. Mleziva formulates a criterion which is rather close to mine, and he remarks that criteria of this kind are open to an unexpected objection. Whereas one generally would expect that the relation of extensional isomorphism is *not strong enough* to enforce the identity of all those states of affairs of propositions which we intuitively recognize to be identical, Mleziva shows that it runs the risk of being *too strong*. He writes: "We are inclined to say that the sentences 'John is human' and 'John is human and John is human' speak about the same state of affairs. But they are not extensionally isomorph.".[21]

We are therefore faced with a dilemma formulated by Mleziva with particular clarity: The question is as follows: 'what extension of the relation of extensional isomorphism is better?

The main problem with the last criterion which I have just examined is however that it applies to a regimented language which is somehow *arbitrary*. My criterion thus lies open to the objection Quine evinced against a similar attempt "If the positing of propositions as objects is serious, any such arbitrarily assembled groundwork for propositional identity must be seen as gratuitous".[22]

7. VANDERVEKEN'S CRITERION *

By availing himself of several concepts borrowed from the theory of speech acts, Vanderveken made decisive progress in the search for a criterion of identity of propositions. The notion he proposes as a criterion is that of 'relation of illocutionary congruence'. The relation of illocutionary congruence itself is defined in terms of the relation of commitment between illocutionary acts which has to be defined first.

In their forthcoming book *Foundations of Illocutionary Logic*, Searle and Vanderveken write:

We shall say that *illocutionary acts* $F(P), ..., F_n(P_n)$ *commit the speaker to the illocutionary act* $F(P)$ [symbolically $F_1(P_1), ..., F_n(P_n) \vartriangle F(P)$] iff in all possible contexts of use i where a speaker a_i succeeds in performing simultaneously illocutionary acts

$F_1(P_1), ..., F_n(P_n)$, (1) he achieves the illocutionary point of F on proposition P with the characteristic mode of achievement and degree of strength of F,(2) he presupposes that the preparatory conditions of $F(P)$ obtain, (3) he expresses with the required degrees of strength the psychological state of $F(P)$, and (4) proposition P satisfies the propositional content conditions of F with respect to i.

For instance, predicting that P commits the speaker to asserting P. The mode of achievement of the illocutionary act is the same in the two cases. The degree of strength of F is the same. The preparatory conditions are also the same. In both cases the speaker is required to have reasons for the truth of the propositional content he asserts or predicts. The sincerity conditions are the same: by performing both acts, he evinces a belief in P. The propositional content in the same conditions, however, are not the same. In a prediction, but not in an assertion, the propositional content has to describe an event posterior to the act of prediction.

When a sequence of illocutionary acts commits a speaker to another sequence of illocutionary acts and conversely there is between the two series of acts a relation of *illocutionary congruence*. The next step is the definition of the identity of illocutionary forces: "Two illocutionary forces F_1, F_2 are identical when all pairs of illocutionary acts of form $F_1(P), F_2(P)$ are illocutionary congruent, i.e. involve exactly the same illocutionary commitments. With this apparatus, propositional identity can be defined in this way: "Two propositions P_1, P_2 are identical when for all illocutionary force F, if it is the case that illocutionary acts $F(P_1), F(P_2)$ are illocutionary congruent. To put is shortly, $P_1 = P_2$ if and only if for all F, $F(P_1) \simeq F(P_2)$.

This definition has non-trivial but intuitively satisfactory consequences. For instance, it entails that $P = P \vee P$ but that $P \neq P(Q \vee \sim Q)$. To show that the latter equation is false, just think of F as a prediction which requires P to be in the future and take Q as a statement about the past. From the failure of the latter equation, we can see that propositional identity is a stronger relation than strict equivalence. From the success of the former, we can see that it is a weaker relation than sentential identity and this is as it should be.

Another nicety of Vanderveken's criterion is this: it provides an algorithm to check the statements of propositional identity, once it is agreed, and Searle and Vanderveken argue that it is so, — that the set of illocutionary forces is recursive. On this proviso, in order to answer the question "Is P identical with Q?" it suffices to prove that (1) For all primitive illocutionary force F, $F(P) \simeq F(Q)$, (2) For any operation O applied to illocutionary forces, the result F' of the application of O to a force F is such that $F'(P) \simeq F'(Q)$ if $F(P) \simeq F(Q)$.

8. SUPPES' GRADUALISM

In *Congruence of Meaning* (1973), Suppes made a fresh start on the problem of defining identity criteria for proposition. He expounds a conception of synonymy which admits degrees like Goodman's but these are viewed as forming a discrete sequence rather than a continuum. To that extent the objection levelled against Goodman's view of meaning might leave Suppes' view unaffected.

Suppes offers a more sophisticated view than the ones previously considered. Isomorphism comes into play at a deeper level than in the former accounts. Whereas Carnap considered finished sentences as objects endowed with a structure, Suppes sees them as the end products of the production rules and applies the isomorphism requirement to the production-tree.

Suppes considers a context free language generated by production rules and provides that language with a set-theoretic interpretation in the spirit of the Montague semantics alluded to in Chapter VII. Terminal words are assigned reference relative to a model and set-theoretic function are assigned to the production rules in order to account for the contribution of syntax to the reference of the compound expression.

Within that framework, Suppes formulates four definitions of *congruence in meaning* of decreasing strength. To hint at his achievement, let us consider the relation of 'strongly $\mathcal{M}$-congruence' which he defines in these terms:

Let S_1 and S_2 be sentences of the given language, that is, derivable by means of the given grammar of the language. Then S_1 is strongly $\mathcal{M}$-congruent to S_2 if and only if the set of semantic trees of S_1 and S_2 can be made identical with respect to each model structure of $\mathcal{M}$,... by identifying isomorphic trees.[23]

"Let us think of our language", Suppes says, "as containing both elementary parts of French and Russian". According to that definition the sentences (1) and (2) can be counted as congruent but neither of them can be said to be congruent with (3).

(1) The book is red.
(2) Le livre est rouge.
(3) Kniga Krasnaya.

Suppes' criterion of strong congruence is neat but it will not satisfy those philosophers who posit propositions as meanings of sentence since it does not succeed in transcending *all* languages. Admittedly Suppes provides also less stringent criteria of congruence of meanings which come closer to this purpose. They seem, however, too *weak* to be labelled criteria of proposi-

THE IDENTIFICATION CRITERION FOR PROPOSITIONS   145

tional *identity*. The question therefore arises: Are there such things as propositions at all? Are there meanings which remain invariant when one switches from one language to another and which the translator is supposed to carry from the source-language to the target-language?

9. INDETERMINACY OF TRANSLATION

It is Quine's contention that there is no such a thing as translation relations objectively valid for sentences in general (he makes an exception for occasional observational sentences). This is the famous Indeterminacy of Translation thesis.

> Manuals for translating one language into another can be set up in divergent ways, all compatible with the totality of speech dispositions, yet incompatible with one another. In countless places, they will diverge in giving, as their respective translations of a sentence of the one language, sentences of the other language which stand to each other in no plausible sort of equivalence however loose.[24]

The bearing of that thesis on the very idea of propositional identity is clearly brought out in this passage:

> sentences are synonymous that mean the same propositions. We would then have to suppose that among all the alternative systems of analytical hypotheses of translation which are compatible with the totality of dispositions to verbal behavior on the part of the speakers of two languages, some are 'really' right and other wrong on behaviorally inscrutable grounds of propositional identity.[25]

But such a supposition, in Quine's opinion, is misguided. He continues: "The very question of conditions for identity of propositions presents not so much an unsolved problem as a mistaken ideal".

The indeterminacy thesis is more than the *epistemological* claim that we shall *never* know among alternative systems of analytical hypothesis of translation which one is *the right one*. It is an ontological thesis to the effect that there is *no* right one. As he says: "There is no fact of the matter." Quine's nihilistic rejection of a third Realm of Meaning, of Bolzano's Satz an Sich is expressed in vivid terms in *From a Logical View* where he imagines a lexicographer at work in a remote country, struggling with the data of his informant, and trying with the help of hypotheses, to build his lexicon by setting up meaningful correlations between the two idioms.

> Quine writes: The finished lexion,... is a case of *ex pede Herculem*. But there is a difference. In projecting Hercules from the foot we risk error, but we may derive comfort from the fact that there is something to be wrong about. In the case of the lexicon, pending some definition of synonymy, we have no statement of the problem; we have nothing for the lexicographer to be right or wrong about.[26]

In his reply to Chomsky, 1968, Quine resorted to the same kind of *ontological* consideration:

> The point about indeterminacy of translation is that it withstands even all this truth, the whole truth about nature. This is what I mean by saying that, where indeterminacy, of translation applies, there is no real question of right choice, there is no fact of the matter... .[27]

The situation can be summed up in this way: scientific theories are underdetermined by all observations (past, present, future and possible), but among the possible competing theories in physics there is only one which can be true of the distribution of particles which constitutes physical reality. Not so for translation manuals. Here there is no fact of the matter to the extent to which there is no such thing as a Realm of Meanings, no such things as *Satz an Sich*.

Quine's thesis of indeterminacy seems to be threatened by triviality. One is tempted to say that if, as far as translation goes, there is nothing to be right or wrong about, then the divergence which can exist between conflicting translations is a divergence about nothing. Several critiques took that line. Professor Young expresses the objection in this way: "If, in contradistinction to rival scientific theories, rival translations have nothing to disagree *about*", then neither is there anything "for translation to be indeterminate *between*",[28] and Professor Schick shares his misgivings: "the difficulty", he says, "is that one's claim of indeterminacy of translation cannot be made interesting without distinguishing among alternative translations by talking about their respective meanings."[29]

That objection, however, can be answered. If we treat analytic hypotheses along the same lines as a coordinate system in physics, we can understand the way in which the former operate. *Before* we have adopted analytic hypotheses, there is no meaning out there to be captured by the translator. But once the manual of translation has been chosen there is such a thing as meaning.

Meaning is not *found* out there, as a 'furniture of the world', it is projected by the manual of translation. In other words, what the manual of translation does is to project the linguistic habits tied up with the source-language onto the target-language. Quine fully acknowledges this asymmetry in the following passage:

> It is only by ... outright projection of prior linguistic habits that the linguist can find general terms in the native language at all, or having found them, match them with his own... The method of analytical hypotheses is a way of catapulting oneself into the jungle language by the momentum of the home language.[30]

But if the manual of translation transfer "linguistic habits" from the source-language into the target-language, in other words, if it *transfers* linguistic habits from one place to another, can we still maintain that it *creates* something? The answer is yes. Manuals of translation generate meaning in so far as *different* manuals of translation connecting the *same* source-language with the *same* target-language can project onto the latter *different* translations. When that situation occurs, it is obvious that the manual of translation has to be held responsible for the *difference* of meaning which results from its application.

The main reason for depriving of ontological import rational principles like the choice of the simplest system often applied to choose among rival translations is this:

...the simplest mapping of language A into language B followed by the simplest mapping of B into language C does not necessarily give the same mapping of A into C as does the simplest direct mapping of A into C. Similarly, the simplest mapping of A into B followed by the simplest mapping of B into A does not necessarily map every item in A back onto itself.[31] I conclude that Quine's disbelief in objective meanings independent of language is well grounded.

REFERENCES

[1] W. V. O. Quine, *Word and Object*, Wiley, New York, 1960, p. 200.
[2] Quine, *Ibid.*, p. 200.
[3] A. N. Prior, 'Is the Concept of Referential Opacity Really Necessary?', *Acta Philos. Fennica*, 1963, p. 190–191.
[4] B. Russell, *Inquiry into Meaning and Truth*, p. 166.
[5] F. de Saussure, *Cours de Linguistique générale*, 1915, 3d edn. 1962, p. 160.
[6] M. Leroy, 'Le Binarisme, concept moteur de la linguistique', *Mélanges de Linguistique, de philologie et de méthodologie de l'enseignement des langues anciennes, offerts à M. René Fohalle*, J. Duculot (ed.), 1969, p. 6.
[7] N. Goodman, 'On Likeness of Meaning', *Analysis* (1949–50), reprinted in Macdonald M., *Philosophy and Analysis*, Oxford Blackwell, 1954.
[8] N. Goodman, 'On Some Differences about Meaning', *Analysis* (1952–53) reprinted in Macdonald, *Ibid.*
[9] A. Naess, 'Synonymity as Revealed by Intuitic', *Philosophical Review* 66, (1957) 87–93.
[10] M. Pêcheux, *Analyse automatique du discours*, Dunod, Paris, 1969, p. 30.
[11] R. Carnap, *Meaning and Necessity*, Phoenix Books, Chicago, 1956, p. 56.
[12] Carnap, *Ibid.*, p. 25.
[13] A. N. Whitehead and B. Russell, *Principia Mathematica* to* 56, Cambridge University Press, 1962, p. 401.
[14] A. Church, *Introduction to Mathematical Logic*, Princeton, 1956, p. 25.
[15] G. E. M. Anscombe, 'Causality and Extensionality', *Journal of Philosophy* (1969) 152.

[16] A. Church, 'Carnap's Introduction to Semantics', *Philosophical Review* **52**, (1943) 299.
[17] W. V. O. Quine, 'Three Grades of Modal Involvement' (1953), *The Ways of Paradox*, Random House (1966), p. 162.
[18] D. Føllesdal, 'Quine on Modality', in *Synthese* **19**, (1968) 154.
[19] D. Føllesdal, *Ibid.*, p. 180.
[20] E. J. Lemmon, 'Sentences, Statements and Propositions', in B. Williams and A. Montefiore (eds.), in *British Analytical Philosophy*, Routledge and Kegan, London, 1966, p. 103.
[21] M. Mleziva, 'Problem fakt u logicke Semantice', *Theorie a Metoda*, **1**, (1969) 74.
[22] W. V. O. Quine, *Word and Object*, p. 205.
[23] P. Suppes, 'Congruence in Meaning', Presidential address delivered at the Forty-seventh Annual Meeting or the Pacific Division of the American Philosophical Association, 1973, p. 26–27.
[24] W. V. O. Quine, *Word and Object*, p. 27.
[25] W. V. O. Quine, *Ibid.*, p. 206.
[26] W. V. O. Quine, *From a Logical Point of View*, 1953, Harper and Row, 1960, p. 63.
[27] W. V. Quine, 'Replies', *Synthese* (1968) 275.
[28] I. Young, 'Rabbits', *Philosophical Studies* **23**, (1972) 180.
[29] K. Schick, 'Indeterminacy of Translation', *Journal of Philosophy* **69**, (1972) 830.
[30] W. V. O. Quine, *Word and Object*, p. 70.
[31] D. Føllesdal, 'Indeterminacy of translation and underdetermination of the theory of nature', *Dialectica* **27**, (1973) 5.
* I am grateful to Daniel Vanderveken for informing me about the criterion of identity of propositions which can be extracted from the theory developed in J. Searle's and D. Vanderveken's *Foundations of Illocutionary Logic* (forthcoming, Cambridge University Press).

CHAPTER X

PROPOSITIONS AND INDIRECT DISCOURSE

1. THE NOTION OF PROPOSITION AND OF INDIRECT DISCOURSE

For a long while the pioneers of contemporary symbolic logic concentrated their research on formal reasonings whose only logical constants were the traditional connectives 'if ... then', 'and', 'not', 'or' and the quantifiers. These are, furthermore, the only notions which are required to analyse a mathematical proof, with the proviso that the '∈' of set theory is added to the list of constants.

For some time, and especially since the publication of Hintikka's *Knowledge and Belief* (1962),[1] logicians have extended their research to include arguments in which the terms 'says that', 'believes that', 'knows that', etc., are part of the logical vocabulary. These investigations had already been initiated with Frege's famous article 'Ueber Sinn und Bedeutung' (1893).[2] Within the framework of this research there are technical reasons for introducing the concept of *proposition*. It is therefore important that we examine these in order to see whether a nominalist theory of the proposition does not conflict with these recent developments in logic.

We shall see that this preoccupation has dominated the reflections of contemporary logicians with a nominalist or at least extensionalist inclination. These logicians have tried to account for the peculiarities in the logic of indirect discourse without appealing to propositional entities. To this extent they had to *represent* these peculiarities as exceptions to extensional logic, which nevertheless can be accommodated within that framework, granting that certain adjustments are made.

The inferences which seemed not to fit into the framework of extensional logic, which was taken as a paradigm for all logic, were viewed in the way that perturbations have been regarded in relation to a general law of celestial mechanics. We must ask ourselves if this solution is not destructive to science, and if it does not impose on further research the impediments of a *doctrinal nominalism* that differs *toto coelo* from the methodological nominalism to which we have subscribed.

This is the conclusion which can be educed from a critical examination of these attempts. We shall, however, see that a scientifically satisfying answer

150   CHAPTER X

to the various problems in the logic of indirect discourse can be found that fully safeguards our previous views, that is, is completely compatible with extensionalism. But this answer requires more than simply filling in open breaches in classical extensionalism with epistemic logic, it requires a thorough reorganization of extensionalism in terms of which classical extensionalism will not longer appear as a norm, but rather as an exception, as an *exceptionally* simple case, something like the hydrogen spectrum which appears today as an impoverished and somewhat degenerate spectrum.

The intellectual mutation which this reorganization requires in us is of the kind Bachelard so lucidly describes in *Le nouvel esprit scientifique* (1939): "Sooner or later the notion of perturbation must be eliminated. One must no longer speak of simple laws which have perturbations, but of complex laws which, in occasional circumstances, admit of certain simplifications".[3]

In our comparison of rival solutions, considerations of a *methodological order* will be dominant. We shall see that the requirement of the systematic compatibility of a theory with the set of facts it accounts for allows us to exercise a choice among the several solutions offered for our assent. Firstly we must construct a logic of indirect discourse that will allow us to exclude illegitimate inferences and, *at the same time*, will guarantee as many intuitively valid inferences as possible. Further, it is necessary that the logician does not presuppose the opposite of what linguists have already established.

2. THE SYNTACTIC APPROACH TO THE PROBLEM OF INTENSIONAL CONTEXTS

In the *Tractatus*, Wittgenstein defends the thesis of the universal applicability of the *principle of extensionality* (and even of the *principle of truth-functionality*) with the same intransigence as the Pythagoreans displayed in their defense of the thesis of the *commensurability* of all numerical relations. These radical positions turned out to be equally untenable. Everyone today admits that the principle of extensionality, and in particular the principle of truth-functionality, admits numerous exceptions. The most frequently cited exceptions are *modal* and *belief* contexts. From

(1)   It is necessary that $2 + 2 = 4$

and

(2)   '$2 + 2 = 4$' is materially equivalent to 'the Earth is round' one cannot infer

(3) It is necessary that the Earth is round.

In other terms, the following inference is illegitimate:

> *Nec. p*
> $p \equiv q$
> *Nec. q*

Likewise, from

(4) Ptolemy believed that $2 + 2 = 4$

and

(5) '$2 + 2 = 4$' is materially equivalent to 'the Earth revolves around the sun'

one cannot deduce:

(6) Ptolemy believed that the Earth revolves around the sun.

To account for these exceptions to the principle of extensionality, several tactics have been used. One may conveniently classify them with the help of Morris' terminology into *syntactic, semantic* and *pragmatic* approaches.

One of the most usual syntactic methods consists in replacing the principle of extensionality by a stricter principle, the principle of intensionality. One can in this way avoid the illegitimate modal inferences, such as that in which we deduce (3) from (1) and (2), yet preserve legitimate modal inferences of the kind

(1) Necessarily $2 + 2 = 4$
(7) Necessarily $2 + 2 = 4$ if and only if $573 + 982 = 1555$

therefore:

(8) Necessarily $573 + 982 = 1555$

For belief contexts the replacement of the principle of extensionality by the principle of intensionality does not suffice. John may believes a sentence '*p*'. This sentence may be strictly equivalent to '*q*', and John may, however, not believe '*q*'.

To resolve this problem in epistemic logic, Carnap proposed a *parallel* solution to that adopted in the case of modal contexts. He suggests replacing the principle of intensionality by a still stricter principle, which one might call the principle of intensional isomorphism. We discussed this in the preceding chapter. Carnap hoped to exclude by means of this principle such illegitimate inferences as that of the above example. Further, and it is mainly for this reason that we are interested in this attempt, he thought he could economize and spare the notion of proposition, analysing

> John believes that the Earth is round

as:

> John gives an affirmative response to 'The Earth is round' understood as an English sentence.

In both points his hopes failed.

As B. Mates[4] and other have shown, the constraints imposed by intensional isomorphism are not sufficient to avoid certain illegitimate inference. Thus, from:

(9)   Whoever believes that all Greeks are Greeks believes that all Greeks are Greeks.

the principle of interchangeability of intensionally isomorphic sentences premits us to *deduce*:

(10)   Whoever believes that all Greeks are Greeks believes that all Greeks are Hellenes

as the sentences 'All Greeks are Greeks' and 'All Greeks are Hellenes' are intensionally isomorphic. Now, it is clear that the sentence (9) can be true and the statement (10) false, unless one takes 'to believe' in a *normative* and *de re* sense rather than in a *descriptive* sense. But in this case one would have to renounce any pretension of having exposed the logical syntax of the verb 'to believe' taken in its ordinary sense, and one would have to consent to disassociating belief from its behavioural manifestations. Carnap resigns himself to this and recognizes that one can no longer regard "an affirmative-response to 'D' as a conclusive indication of belief in D".[5]

On the other hand, the reduction of propositions introduced by 'that' to the sentences in question, as proposed by Carnap, has raised the following objection. It is formulated by Church in 'On Carnap's Analysis of Statements of Assertion and Belief' (1950).[6] According to Carnap, the sentence:

(11)   Seneca said that man is a rational animal

may be analysed as:

(12)   Seneca uttered a sentence meaning 'man is a rational animal'.

But, as Church points out, when these two sentences are translated into another language, the sentence between the quotation marks in (12) remains intact, that is, it is not translated. Heeding this, if (11) and (12) are translated into French, a French speaker with no knowledge of English will understand (11) but not (12).

This translation test reveals, therefore, that the *analysans* (12) does not have the same sense as the *analysandum* (11); it changes the informative content.

In other terms, (12) is not *analytically* equivalent to (11). For someone who rejects the analytic-synthetic distinction, as Quine does, Church's objection is not decisive, but it is devastating against Carnap who accepts this distinction.

In 'An Inscriptional Approach of Belief Sentences' (1954), I. Scheffler proposed a solution different from Carnap's analysis. This is a hypernominalistic solution designed to escape from the objections levelled against Carnap's attempted reduction.

Inspired by a suggestion of Goodman's, which he develops in an original and ingenious way, Scheffler proposes "construing that-clauses ... as single predicates of concrete inscriptions, and taking every inscription denoted by a given that-clause as a *rephrasal* of every other so denoted".[7]

Church's sentence (11) now becomes:

(13) $(Ex)(Ey)(x = \text{Seneca} \cdot \text{that-man-is-a-rational-animal } y \cdot \text{Inscribes } xy)$

This analysis of (11) is protected from the numerous objections raised by Church against Carnap's analysis, of which we considered the most famous. The inscription predicate, that is 'that-man-is-a-rational-animal y', can indeed be translated on the same basis as 'x = Seneca'. On the other hand, it does not risk the ire of the nominalists, as we have here with the inscription predicate a predicate of concrete expressions and not at all a name of an abstract entity such as the proposition. However, this argument will not suffice to justify it in the eyes of those, like ourselves, who refuse to opt from the outset in favor of a doctrinal nominalism.

Despite its subtlety, Scheffler's solution cannot be adopted. One can, in fact, object that it does not extend to neighboring problems, that it is local and limited like the special postulates formerly employed in astronomy in order to save geocentrism.

In *Word and Object* (1960), Quine pointed out the limits of Scheffler's solution:

154   CHAPTER X

When Scheffler extends his method to idioms of propositional attitude other than indirect quotation, however, a peculiar difficulty does arise: how are we to say e.g. that Paul believes something that Elmer does not? It will not do to say that Paul believes-true some utterance that Elmer does not believe-true, for it may happen that no such utterance exists or ever will... .[8]

3. PRIOR'S NOMINALIST SYNTAX

Another nominalistic solution to the problem raised by intensional contexts and which one may identify with the syntactic approach is that offered by Prior in *Objects of Thought* (1971). It is a far-reaching theory which we must examine carefully.

Prior does not attempt to reduce propositions to sentences. He tries, however, to rid himself of propositions by showing that they are logical constructions, as Russell had tried to rid himself of classes by treating class-names as incomplete symbols. Russell's attempt failed. As Quine has shown, Russell succeeded in freeing himself of classes only by surreptitiously having recourse to other Platonic entities: attributes. Is Prior more successful than Russell in his undertaking? We shall try to show that he is not. Very serious objections may be urged against Prior's theory.

Prior asks us to abandon the usual logical analysis of 'X says that p', 'X believes that p', etc., according to which 'believes' is a dyadic predicate which applies to ordered pairs, the first term of which is a thinking subject and the second a propositional entity symbolically represented by the expression 'that p'. In place of this, he proposes to treat 'says that' or 'thinks that' as unitary expressions whose syntactic status is not that of a predicate but of a hybrid entity: "They are as it were predicates at one end and connectives at the other".[9]

Since Prior further admits that "Only names designate objects, sentences do not, and verbs do not",[10] the temptation to posit propositional entities as designata of expressions like 'that grass is green' disappears.

In Prior's theory, general sentences such as 'There is something that Paul believes and that Elmer does not believe' that cause Scheffler some difficulty, receive an analysis which appears satisfactory *within certain limits*. We have already said something about this in Chapter VI, but it is important to return to this point in the light of the more articulated information contain in *Objects of Thought*.

Since 'x believes that p' is treated as a sentential connective on its right side, the variable 'p' syntactically plays the role of a sentence and not of a name. This distinction is important: "I cannot see anything ... that need compel us to treat variables that do *not* stand for names of objects as if they did".[11] Since only nominal variables standing for names, like x in 'x believes that p', can have values which are extra-linguistic entities, propositional variables like p can only have substituends, and these substituends will obviously be sentences. Therefore, Prior can formalize the above-quoted sentence

(14)   There is something that Paul believes and that Elmer does not believe,

in the following way:

(15) $(\exists p)$(Paul believes p · Elmer does not believe p)

By writing such a sentence, one is not at all committed to an ontology including propositions if one adopts an analysis which combines an interpretation of 'believes that' as a connective with a substitutional interpretation of quantification. This is what Prior does.

Prior goes even further and *uselessly* involves himself in an unacceptable analysis when he writes that 'X believes that grass is pink' must be analysed into 'Grass is, in X's opinion, pink'.[12] When one applies this reduction to iterated beliefs such as 'X believes that Y believes that Z believes that grass is pink' one obtains an unintelligible result. But this part of Prior's theory may be omitted without damaging the remainder of the theory.

As we shall see, Prior succeeded in avoiding propositions in numerous circumstances where others were constrained to admit them. He was, however, not successful everywhere, and even those of his views which are satisfying when taken alone, cannot be retained if we wish to achieve a unified theory.

At the beginning of this chapter we saw that arguments involving belief neither conform to the ordinary principle of extensionality nor to the principle of extensionality reinforced by strict implication nor even by intensional isomorphism. The fact that two sentences are strictly equivalent and isomorphic does not suffice to *allow* interchangeability. Prior proposes, therefore, to strengthen the principle by introducing a stricter connective. In 'Is the Concept of Referential Opacity Really Necessary?' (1963),[13] he maintains that for two sentences 'p' and 'q' to be interchangeable *salva veritate* in a belief context, it is necessary that p express the same proposition as q.

One might feel that here we have just lost all of the ground we had gained from those who think that it is necessary to appeal to propositions. This is not so. The word 'proposition' does appear in Prior's formulation of the interchangeability principle, but it does not play here the role of the *name* of an entity, but that of a *fragment* of a syntactic connective, just as does the word 'if' in the connective '*p* if and only if *q*'.

By maintaining that 'that *p* is the same proposition as *q*' is a connective and not a predicate, Prior avoids the objection of reification, but he is open to other criticisms. For a connective to have a sense, one must provide syntactic and semantic rules governing its use. On this point Prior restricts himself to stating two fundamental axioms which constitute an implicit definition of 'is the same proposition as' (symbolized by '*I*'); namely, *Ipp* and *CIpqIδpδq*. 'δ' being used as a variable standing for expressions which form a sentence out of a sentence. As L. J. Cohen remarks in "Critical Notice" (1973),

> He does not give any semantics for '*I*', i.e. any truth-rules for sentences of the form '*Ipq*'. In such a semantics he might have specified what kinds of verbal transformation a sentence can undergo without losing its propositional identity, and then at least we should have had a theory to check.[14]

Cohen's objections appear to us to be completely justified. Prior says, in effect, that when one concentrates on content rather than form, the proposition that *Jones asks himself whether all bachelors are not married* is exactly the same proposition as *Jones asks himself whether all unmarried men are unmarried*. From this he seems to admit implicitly that two sentences of different form can nevertheless express the same proposition. It is exactly at this point that Prior makes a questionable appeal to the concept of a proposition, not because he reifies propositions — he does not do that — but because he supposes a problem to be solved which is not solved and perhaps never will be solved, namely, the problem of a criterion of identity for propositions.

4. L. J. COHEN'S EXTENSIONALIST SYNTAX

In *The Diversity of Meaning* (1962; revised edition 1966), Cohen adopts Prior's point of departure; for Prior non-extensionality is "the product of statement-forming operators on statements rather than of words being used in some non-customary sense". In other words, he believes that non-extensionality has its source in syntax rather than in semantics.

Cohen's original contribution consists in analysing these operators and in reformulating sentences in which they occur into sentences for which the principle of extensionality holds. This *extensional reformulation* is illustrated by the following example: "The sentence

A Cretan says that it is raining

becomes

$$(\exists x)(Cx \circ Rx)(Tx \equiv p)$$

where 'C' stands for 'is a statement by a Cretan', 'R' for 'is a statement that it is raining', 'T' for 'is true', 'p' for 'it is raining', and individual letters include sentence-tokens in their domain of reference".[15]

This extensional device for formalizing arguments containing operators such as 'believes', 'says', 'doubts', etc., poses certain problems. For example, how would Cohen account for the non-validity of the following argument:

(16)   1. George IV doubted that Scott = the author of Waverley
        2. Scott = the author of Waverley
        3. George IV doubted that Scott = Scott.

In 'On a nominalistic analysis of non-extensional contexts' (1972), Simpson raises this question and claims that Cohen must choose between two equally unsatisfactory answers. The first consists in saying that the logical form of this invalid argument is not that of the valid schema

(17) $x = y$
        ϕx
        ϕy

but rather that of the invalid schema

(18) $x = y$
        ϕp
        ϕq

Simpson offers the following objection to this solution: "This implies that the 'that'-clauses are not analyzable here".[16] He considers it arbitrary to deny that 'Scott' and 'the author of Waverley' are subjects of (1) and (2), respectively.

According to the second possibility of accounting for the non-validity of (18),

... the true logical form of (1) and (2) is such that their subordinate clauses only appear as parts of a predicate term applicable to tokens; thus, (1) is transformed in something like

(1')   There is an x such that x is *a saying that George IV doubted that Scott = the author of Waverley*, etc.

where the underlined phrase stands for a property attributed to the token x. So, it would not be here any possible substitution position for 'Scott' and 'the author of Waverley'.[17]

But again, Simpson protests against Cohen's arbitrary decision that these predicates are unanalysable.

Of course, Mr. Cohen may *rule* that in his formalized language such predicates are non-analysable units, and this would be all right; but it might hardly be considered as a clarification of a problem originating from the ordinary language.[18]

It seems, therefore, that a syntactic approach, even if it is as elaborate and subtle as Cohen's, cannot account for the logic of intensional contexts. Let us see if the semantic approach is more effective.

5. FREGE'S DUALIST SEMANTICS AND EPISTEMIC LOGIC

In 'Ueber sinn und Bedeutung' (1892), Frege initiated modern research in this area. The thesis of this classic essay is well-known. Wishing to explain how an identity statement such as 'the morning star = the evening star' can be true and non-trivial, Frege distinguishes two dimensions of meaning: *sense* and *reference*. In order to solve the problem of non-trivial identity, it suffices to note that the expressions on either side of the '=' sign may be coreferential without being synonymous. Also note that the distinction between sense and reference can be extended to sentences. This is not surprising since, for Frege, sentences are names. The reference of a sentence is its truth-value, the sense of a sentence is the proposition it expresses.

So far there is nothing in Frege's analysis to which a nominalist could object. The nominalist can, in fact, coherently distinguish *sense* from *reference*, or *intension* from *extension*, as long as sense or intension are not posited as autonomous entities, are not reified.

But Frege developed his theory and made later application of his distinction between sense and reference which, according to some, does not enjoy the same philosophical neutrality as it might be thought to possess in its application to the identity puzzle. More precisely, these are the problems raised by indirect discourse and intensional contexts. According to Frege, "...in indirect discourse, [words] are used *indirectly*, or have *indirect* nominata".[19] The same applies to sense.

What is the *indirect referent* of a word? Frege answers: "In indirect discourse ... the indirect (oblique) referents ... coincide with the sense". In other words, in indirect discourse the reference of a word is its ordinary *sense*, it is named by the word.

Now, we know that naming and reification go together. To regard sense as the *nominatum* of a name is to treat it as an independent entity. Therefore, the sentence *p* in indirect discourse as '*A* believes that *p*' has as its referent not a truth-value but a *proposition*. At precisely this point Frege's doctrine becomes tainted with propositional Platonism. The proposition ceases to be the *sense* of a sentence in order to become its *referent*, the *nominatum* of the sentence.

Frege's 'propositional Platonism' is certainly not *gratuitous*. Thanks to his conception of oblique reference, Frege provides us with an explanation of the non-applicability of the principle of extensionality in belief contexts. This explanation, on the other hand, is not *ad hoc*. It resulted from a conceptual distinction which was *independently* required to resolve the identity puzzle.

Frege's theory results in a two-head logic, which was subsequently developed by Carnap; it requires principles of the substitutivity of identicals which vary with context. In an article 'Sobre la eliminacion de los contextos oblicuos' (1967),[20] Simpson has shown that it is possible to construe Frege's theory in such a way that the same principle of substitution (Leibniz' law of the substitutivity of co-referential terms) is applicable to both direct and oblique contexts. Unfortunately, this remarkable attempt at unification can only be applied to certain quantified sentences.

6. CARNAP'S DUALIST SEMANTICS

Carnap and Church took up Frege's views and have elaborated them in an original way. We shall restrict ourselves here to a critical examination of Carnap's theory, the most recent version of which appears in his 'Replies and Systematic Expositions' at the end of the monumental work edited by Schilpp in 1963.[21] Carnap, in fact, *juxtaposes* a logic of extension with a logic of sense, both constructed on the same pattern. These two parallel logics are mainly contrasted by the following traits characterizing object-languages for which they provide the rules of inference.

(a) The extensional language contains variables whose values are extensions (such as individuals, classes and truth-values).

(b) In order to combine statements or other designators (names, predicates), it contains a binary connective such that every statement in which it occurs as the main connective will be true if and only if the expressions which it combines are co-extensive.

(c) The extensional language also contains an interchangeability principle according to which co-extensive expressions are interchangeable *salva veritate* in all contexts.

The language of sense, on the other hand, satisfies the following conditions:

(a') It contains variables of sense, that is, variables whose values are senses.

(b') It contains a binary connective such that every statement in which it occurs as the main connective will be true if and only if the expressions which it combines are synonymous.

(c') Finally, it contains an interchangeability principle according to which synonymous designators are interchangeable *salva veritate* in all contexts.

One will note that in condition (a') Carnap makes an ontological commitment to propositions. In order to appreciate the merits of this theory, one must see how it works, when applied to arguments for whose analysis it was constructed; we shall also apply it to other cases within its domain, cases which it seems were not foreseen.

Consider the following two arguments:

(19)   Hesperus is smaller than the sun
       Phosphorus is identical with Hesperus
       Therefore, Phosphorus is smaller than the sun.
(20)   Albert believes that Hesperus is smaller than the sun
       Phosphorus is identical with Hesperus
       Therefore, Albert believes that Phosphorus is smaller than the sun.

The first argument is of a sort which occurs in elementary mathematics. It is valid, and it is easy to account for its validity within the framework of extensional logic. The second is not valid. If Albert knows no astronomy, he may not know that the morning star or, more precisely, the morning planet, Phosphorus, is the same as the evening star, Hesperus. The premises are, therefore, true and the conclusion false. Now, it is easy to account for this non-validity if one appeals to the logic of sense, which allows us to interchange the words 'Hesperus' and 'Phosphorus' if they are synonymous. They are not synonymous, and the premise affirming the identity of Hesperus and Phosphorus does not assert such a synonymy as it affirms a relation between bodies and not between senses. Each of these arguments thus comes under the jurisdiction of a different logic.

We have just seen that these two logics have their own semantics — the value-range of their variables differ, taking either extensions or senses as values — and their own syntax: the connective of extensional equivalence in the one case and that of synonymy in the other, and the principle of interchangeability governing each of them. From the outset, one might expect that this two-headed logic would be badly adapted to *mixed* arguments, that is, to inferences which simultaneously involve extensional and sense logic. This kind of argument will serve as a testing ground to judge the merits of the two types of solutions which we proposed to examine here. It is not difficult to construct an argument of this kind. The one we are going to examine involves an application of the rule of existential generalization.

(21)   Albert believes that he is pursued by Bernard and he is in fact pursued by Bernard.
Therefore, there exists an individual by whom Albert believes he is pursued and by whom he is in fact pursued.

If we wished to formalize this argument in the usual way, we would write, letting 'B' represent 'believes' and 'Ex' represent 'there exists'

(22) $Ba(Pba) \cdot Pba$
∴ $(Ex) [Ba(Pxa) \cdot Pxa]$

Unfortunately, it is not possible to adopt this formalization without giving the variable 'x' an *equivocal* interpretation. Indeed, conforming to Carnap's theory, this variable will admits *senses* as values in its first occurrence and *individuals* as values in its second occurrence. To avoid this ambiguity, one would have to adopt two kinds of variables, but then the initial quantifier could not bind both variables and remain a well-formed expression. It seems, therefore, that Carnap's dualistic logic is not only incapable of explaining the validity of the above argument, but even of representing it symbolically in the formal language.

Certainly the possibility of developing Carnap's theory so that it escapes these criticisms is not excluded, but it is a task which remains to be done.

Carnap begins by saying that it is possible to translate intensional language (which, in his terminology, is a language intermediary between an extensional language and a sense language) into extensional language. The possibility of this translation depends upon the existence of a biunique correspondence between the values of the extensional variables and certain concepts representing intensions, concepts he calls quasi-intensions.

The quasi-intensions of n are themselves extensions, but of a higher type level than n — which lets us glimpse the technical complexity that a solution to our problem within the framework of Carnap's theory will present.

7. QUINE'S UNITARY EXTENSIONALISM

We saw that Frege contrasted direct with oblique contexts. The use of the word 'oblique' is very judicious. It evokes the idea of deviation and this is exactly what we were concerned with. For Frege, indirect discourse, *oratio obliqua*, causes reference to *deviate*; a sentence introduced by 'says that' takes as its reference what in direct discourse was its *sense*, that is, a proposition.

Quine's diagnosis of the logical irregularities associated with the contexts traditionally called intensional differs greatly from Frege's. Instead of saying that in these contexts reference deviates, i.e. it is displaced from truth-values to propositions, Quine says that reference is *obfuscated* and *disappears*. Attitudes' (1956),[22] he pointed out the effect of these contexts on the application of the rule of existential generalization.

Quine's diagnosis of the logical irregularities associated with the contexts traditionally called intensional differs greatly from Frege's. Instead of saying that in these contexts reference deviates, i.e., it is displaced from truth-values to propositions, Quine says that reference is *obfuscated* and *disappears*. In place of the distinction between *direct* and *oblique contexts* he substitutes the distinction between *transparent* (an expression he claims to have borrowed from Russell) and *opaque contexts*.

The paradigm of referential opacity is quoted discourse. According to the usual view, quotation marks around a sentence form a name of that sentence. As a name is the smallest unit of sense, it is not divisible, hence, *It is raining* is no more a part of 'It is raining' than 'rat' is a part of 'Socrates'. It is a surprise to no one that it is forbidden to substitute co-referential terms within a name or to quantify over part of a name in the following way:

$\quad (\exists x)$(xenophobia).

Quine *assimilates* oblique contexts to opaque contexts. He certainly does not claim that 'He believes that p' is synonymous with 'He believes that 'p'', which would expose him to the criticisms levelled at Carnap, and which would oblige him to speak of synonymy, whereas he declines to do so. What he is claiming is that oblique contexts alter reference in the same way that quotation does; they obscure and suppress it. This assimilation allows Quine to account for the non-validity of certain arguments, those that rest upon an application of the principle of substitutivity or of the rule of existential generalization to belief contexts, without having to admit, like Frege, propositional entities as the referents of sentences introduced by 'Believes that', 'says that', etc.

Consider, for example, the following inferences:

(23)   Tom believes that Cicero denounced Catiline
            Cicero = Tully

Therefore, Tom believes that Tully denounced Catiline.

(24) Tom believes that Cicero denounced Catiline
Therefore, $(\exists x)$ (Tom believes that x denounced Catiline).

The reason why these arguments are not valid becomes evident if one assimilates them to the following arguments where the predicate 'believes-true' which I borrow from Quine stands for a relation between a believer and a sentence.

(25) Tom believes true in English 'Cicero denounced Catiline'
Cicero = Tully
Therefore, Tom believes true in English 'Tully denounced Catiline'.

(26) Tom believes true in English 'Cicero denounced Catiline'
Therefore, $(\exists x)$ (Tom believes true in English 'x denounced Catiline').

If it were limited to this, Quine's theory would be too simple. It would explain the non-validity of certain inferences, but would leave unexplained and even make inexplicable the validity of others. Now, a good theory must explain *both* phenomena.

Consider the following argument where the Quinean predicate 'believes to be satisfied' is construed on the model of 'believes-true'.

(27) Of Cicero, Tom believes that he denounced Catiline
$(\exists x)$ (Tom believes the propositional function 'y denounced Catiline' to be satisfied by x).

This inference is perfectly valid. It does not require giving up extensional logic and, furthermore, its premise is obtained by a syntactic transformation on the premise of (23), a transformation which preserves one of the two possible senses of that premise. How, if one assimilates oblique to opaque contexts, can one explain the validity of this inference? Quine replies as follows: opacity and transparency are not intrinsic properties of words, but *properties relative* to sentential occurrence. In (27) the occurrence of 'Cicero' and the last occurrence of the variable 'x' are outside of an opaque *zone* and, hence, the inference is valid, whereas (26) is not.

This distinction between an opaque and transparent *position*, as Quine remarks, is inscribed in grammar. In 'Nouns and Noun Phrases' (1968), the linguist E. Bach joins Quine and detects an ambiguity in the sentence:

(28) She wants to marry a man with a big bank account.

which he explains in terms of two deep structures which underlie the same surface structure:[23]

(29) There is a man with a big bank account that she wants to marry.
(30) She wants there to be a man with a big bank account and that she marry him.

Parallel to the distinction between opaque and transparent position, Quine cites another. He distinguishes two senses of the verb 'to believe', an *intensional* sense, where contexts introduced by this term are not amenable to existential generalization and interchangeability of co-referential terms, and a *relational* sense which permits inferences of both kinds. In the first case, the verb 'to believe' is analysed as a *dyadic* predicate relating individuals to sentences, $R(x, \ 'p')$. In the second case, it is analysed as an irreducible *triadic* predicate relating two individuals to a propositional function, $R(x, z, \ '\phi y')$.

The inferences which appear to be peculiar to the logic of indirect discourse and of belief are, therefore, either reduced to ordinary inferences of classic extensional logic, thanks to a grammatical transformation, or outlawed, frozen as it were by their assimilation to quotation. Since extensional semantics is the only one needed and this does not appeal to propositional entities, Quine's views are attractive for nominalists. But doesn't this elimination of intensional logic entail an impoverishment of theory and make it incapable of accounting for the facts?

Unhappily, the answer to this question is affirmative. Quine's extensional logic is not able to account for certain intuitively sound 'mixed' arguments without deforming them. Let us return to the mixed argument (21). Contrary to the case with Frege and Carnap, Quine would be able to account for the validity of this argument, but in order to do so he would have to adopt the relational interpretation of 'believes' and sacrifice the intensional interpretation. This can be safely done in the case of the argument in question, but there are others where this transformation would be mutilating.

Consider the following sentence, which is analysed by Brian Loar in 'Reference and Propositional Attitudes' (1972):

(32) Michael thinks that that masked man is a diplomat but he is not.

The expressions 'that masked man' occurs here, as in our example (21), both within and outside of an oblique context. In order to correctly apply existential generalization one would, thus, have to begin by exporting the referring expression with respect to which the generalization is made to a position outside of the opaque context. We might attempt something like:

(33) Of that masked man, Michael thinks that he is a diplomat but he is not.

Unfortunately, this transformation completely alters the sense of the premise. In fact, it is not of a certain individual under any description that Michael thinks is a diplomat, but of an individual identified by the expression 'the masked man'. If the individual had not been masked, Michael would no doubt have recognized him and would not have mistaken him for a diplomat. Exporting a constant from an opaque zone to a transparent zone, an operation which for Quine must precede a correct application of the rule of existential generalization, has as one can see a double effect. It confers a referential value on the sign, which was tolerable in the case of the mixed argument (21), where the sign had a reference in its second occurrence anyway; but, besides this, exportation destroys the descriptive value of the sign, and this, as Loar's example shows, is much more serious.

Loar proposes an analysis of mixed recurrences that allows us to preserve the advantages of exportation without the inconveniences. Exploiting the distinction between attributive and referential uses of referring expressions introduced by Donnellan, Loar assigns to singular terms a 'dual contribution' to the truth-conditions of the sentences containing them.

The logical form of the sentence

(34) Michael believes that that masked man is a diplomat

is neither

(35) Michaels believes 'that masked man is a diplomat'

nor

(36) Michael believes of that masked man 'x is a diplomat'

but

(37) Michael believes of that masked man 'x is a masked man and x is a diplomat'.

The interpretation of 'that masked man' is, therefore, simultaneously relational and intensional. It is not simply relational because 'something stronger is being asserted — something whose truth depends upon Michael's *identifying* the man as such and such and believing of him *under* that description that he is a diplomat'.[24] This analysis solves our problems because it allows us to quantify over the transparent occurrence of 'that masked man' without thereby altering the sense of the premise.

However, Loar's solution is still not adequate. It only permits us to handle sentences in which referring expressions but not predicates play a double role, but cannot deal with the following inference:

(38) John believes that Ralph is writing and Ralph is not writing. Therefore there is something which John believes Ralph to be doing and which Ralph is not doing.

Indeed, the premise would be written as:

(39) Of Ralph, John believes-true '*x is Ralph and x is writing*' and of Ralph, '*x is writing*' is not true.

Now, no one would even consider substituting the *same* metalinguistic variable for the two underlined predicates since the two underlined expressions are obviously different, therefore, one could not infer:

(40) $(\exists \phi)$ (Of Ralph, John believes-true ϕ and ϕ is not true).

In order to resolve *systematically* all of these problems, one must have recourse to a general theory of broad scope. Hintikka's plural extensionalism, at the present state of understanding, seems to me the most systematic and satisfying answer to the multiple problems one has to face in elaborating an epistemic logic.

8. CRITICISMS ADDRESSED TO QUINE'S NOMINALIST THEORY: KAPLAN'S ALTERNATIVE SOLUTION

In 'Quantifying In' (1968),[25] Kaplan formulated, in opposition to Quine's analyses, a number of constructive objections. We only mention here those which are relevant to our subject.

The first objection is *methodological* in nature. In order to deal with the complexity of the logic of belief, Quine must introduce into this logic a *primitive predicate*, namely the verb 'believes that' in the relational sense of belief. His solution, although it permits the simultaneous solution of several problems and is thus not *ad hoc*, is nevertheless *local*. It requires the introduction of analogous distinctions for other predicates.

The second objection, similar to the first, is the following: in reducing the sentences of indirect discourse to the sentences in question, which constitute indivisible totalities, one is not able to explain how the meaning of this totality is a *function* of the meaning of its parts. And from this, arises the psychological problem of explaining how one can learn a natural language containing an infinite number of primitive predicates, a consequence underlined by Montague in 'On the Nature of Certain Philosophical Entities' (1969).[26]

Now, it is incontestable that the meaning of 'Galileo believed that the Earth is round' is a function of its parts, and depends upon the *meaning* of 'Earth', of 'is', of 'round' in a sense in which the meaning of 'Galileo uttered the words: "The Earth is round"' does not depend on the meaning but only on the *form* of 'Earth', of 'is' and of 'round'. This objection of Kaplan's appears very important to us. As with Montague, it indicates a concern of his not to accept a logical theory which ignores the genetic dimension of language and which *excludes* the generative conception of meaning which we have defended in this work.

In 'On Saying that' (1968), D. Davidson[27] is also sensitive to the same issues. He objects that the reduction to quoted contexts removes them from the scope of Tarski's semantic theory of truth, whose recursive definitions offer us a rigorous account of the combinatorial and generative character of meaning.

Finally, the distinction between the two belief concepts does not suffice to avoid certain illegitimate inferences. Kaplan has shown, by means of an ingenious counter-example, that the type of 'exportation', that is to say of existential generalization from outside an opaque context which Quine tolerates, must admit restrictions not foreseen in his theory.

For all of these reasons, Kaplan proposes a return to Frege whose views he elaborates, but — and this is of utmost importance for us — without accepting Frege's Platonistic conceptions. Moreover, he questions whether Frege is really forced by the logic of his system to adopt the realist ontology that one usually imputes to him.

If one has believed this, it is, according to Kaplan, because one has not heeded Frege's remarks concerning the *role* of quotation and oblique contexts. The accusation of Platonism addressed to Frege is based upon a confusion.

It conflates two separate principles: (a) [the principle affirming that] expressions in oblique contexts don't have their ordinary denotation (which is true), and (b) [the principle affirming that] expressions in oblique contexts denote their ordinary sense (which is not, in general, true).[28]

Frege allows quantification over variables occurring in oblique contexts, but in order to avoid invalid inferences and to preserve valid inferences, he assigns to expressions occurring in *oblique* contexts a *different* reference, a different denotation, than they have in ordinary contexts. What is illegitimate, according to him, is not quantification into *oblique contexts*, but *simultaneous* quantification into oblique and ordinary contexts, for in doing this one uses the variable in an *equivocal* way.

If one takes the necessary precautions against ambiguity, one may, according to Frege, quantify over variables occurring in what Quine today calls opaque contexts. This latter quantification, which we had previously judged absurd, ceases to be so if we make use of an inoffensive supposition, namely, the supposition that the signs within such contexts denote themselves.

Kaplan makes this assumption which he finds suggested by Frege and exploits the possibilities which it opens. He thus allows quantification into quotation contexts, but in order to *avoid equivocation*, he takes unique values for these opaque variables. These values are nothing other than the signs themselves. We remark that from the *nominalist* position we have defended up to now, the ontological assumption that Kaplan requires us to make is perfectly acceptable. It simply requires us to assume the existence of signs.

The second and most important of Kaplan's innovations is the following. After having *separated* the two value-ranges of the variables: the value-range for the transparent variables and that for the opaque variables — we recall that these latter are not variables at all for Quine, but parts of names — Kaplan *systematically connects* these by means of the dyadic predicate 'denotes'.

One cannot stress enough the importance of this innovation. Neither Church nor Frege built a bridge between extensional and intensional logic, consequently the latter stood out like an erratic block, a foreign body, without connection to extensional logic. This gap, this absence of a bridge between the two logics, has been emphasized by Prior, who, in 'Modal Logic in the Style of Frege' (1957), remarks that the system of Church, 'A Formulation of the Logic of Sense and Denotation' (1951), "... contains no law of the form $CLpp$".[29] In other words, it is not possible to deduce 'p' from 'it is necessary that p'.

It is fundamentally this problem of the unification of intensional and extensional logic that Quine encounters when he shows how one may *extract* an expression from an opaque context and place it in a transparent position. But this solution is hardly satisfying because this extraction is extremely onerous; for at every turn he must introduce a new predicate. It is precisely here that Kaplan's innovation applies: Kaplan needs only a *single* predicate, the predicate 'Δ' ('denotes').

Kaplan writes that for Quine:

... to move an expression in an opaque construction to referential position, a new *primitive* predicate ... had to be introduced and supplied with an interpretation. In contrast, the same effect is achieved by Frege's method using only the original predicates plus logical signs, including 'Δ', and of course the ontological decomposition involved in the use of the Frege quotes.[30]

Kaplan's extremely promising theory forces us to renounce Quine's views regarding opaque contexts, but it is important to note that for the nominalist no unacceptable concession is implied by the new theory. One can say as much for the extensionalistic theory of Hintikka, about which we shall now say a few words.

9. HINTIKKA'S PLURALISTIC EXTENSIONALISM

The most embarrassing puzzle with which we have been thus far confronted is that produced by inferences in which the same variable occurs both in an ordinary and in an opaque context. How are we to avoid ambiguity in the variable and at the same time maintain a distinction between contexts?

Hintikka has opened a truly original and fruitful path in this domain. He places back-to-back Frege's *dualistic* solution which opposes a theory of sense to a theory of reference and a *monistic* theory which sacrifices sense for reference, as with Quine, or reference for sense, as with Carnap if we attend to certain passages in his work. Hintikka modifies fundamentally the *initial facts* of the problem, even to the extent of questioning the distinction between sense and reference that was presupposed by the afore-mentioned authors, a presupposition which determined the range of possible solutions. Instead of opposing these two theories as contrary to one-another, Hintikka presents them both as theories of references but treats the (classic) theory of reference as a theory of *simple* reference and the theory of sense as a theory of *multiple* reference.

Thus, Hintikka does not deny the contrast, but treats what was considered as a difference in *nature* as a difference in *degree*. The objects to which referring expressions occurring in modal contexts refer are the same *objects* on an ontological level. In 'Modality as Referential Multiplicity' (1957), which previews his later research, Hintikka writes with respect to the occurrence of variables in modal or epistemic contexts: "They range over exactly the same kind of ordinary individuals as are the values of the variables of the corresponding non-modal statement".[31] The only difference is that these individuals are not necessarily part of the real world, they can be part of a possible, non-realized world.

In 'Meaning as Multiple Reference' (1968),[32] Hintikka brought out with particular clarity the role played by the notion of possible world in the definition of propositional attitudes:

> My basic hypothesis is that an attribution of any propositional attitude to the person in question involves a division of all the possible worlds which we can distinguish in the part of language we use in making the attribution into classes: into those possible worlds which are in accordance with the attitude in question and into those which are incompatible with it.

On might object that people hold incompatible beliefs in which case there is no possible world compatible with them. Hintikka's way out would be to use 'belief' in a normative sense and to consider idealized believers who hold only compatible beliefs. The meaning of the division between the two above-mentioned classes of possible worlds may be brought out e.g. by paraphrasing statements in terms of propositional attitudes by speaking only of the class of all possible worlds compatible with them. The following example will illustrate these approximate paraphrases:

a believes that p = in the worlds compatible with what a believes it is the case that p.

Similarly,

a knows that p = in the worlds compatible with what a knows, it is the case that p.

Once these preliminary hypotheses are admitted, all the rules of ordinary logic which one applies to oblique contexts can be accounted for. These abnormal inferences are legitimized.

For example, the shortcomings of the co-reference of terms as a condition on interchangeability in contexts expressing knowledge or belief will be explained by the fact that terms which are co-referential in the actual world are not necessarily co-referential in all possible worlds, and, among others, not in all worlds compatible with the knowledge or beliefs of a given person. A remedy has therefore been found. In order to make the following inference valid:

(41) 1. Albert knows that Cicero denounced Catiline

2. Cicero = Tully
3. Therefore, Albert knows that Tully denounced Catiline

one must add a premise:

4. Albert knows who Tully is.

A solution of the same kind is applied in the case of existential generalization. The inferences in epistemic logic whose apparent invalidity caused a problem are in this way reduced to enthymema. Is the logical legality of inferences involving oblique contexts obtained at the price of an *ad hoc* multiplication of auxiliary premises? Hintikka reassures us on this point. In *Models for Modalities* (1969), he writes: "the only possible types of auxiliary premises (uniqueness and/or existence presuppositions) will then be of the following kinds

(42) 1. $(Ex) [x = b \cdot a \text{ knows } (x = b)]$
2. $(Ex) a \text{ knows } (x = b)$
3. $(Ex) (x = b)$

depending on whether we are considering a formula in which 'b' occurs both inside and outside the scopes of '*a* knows' ..., only inside, or only outside".[33]

The way in which Hintikka resolves the logical problems raised by oblique contexts is *technically* superior to others. It accounts for a greater number of 'facts'. Thus, for example, it is capable not only of *formalizing* the mixed inferences over which other theories have stumbled, but it is also capable of stating precisely under what conditions such arguments are valid. In other words, it amounts to a systematic way of recovering deleted premises.

Such an inference as (22) augmented by the premise upon which its validity depends, has the following form:

(43) $[Ba (Pab)] Pba$
$(\exists x) [(x = b) \cdot B a (x = b)]$
Therefore, $(\exists x) \{[B a (P x a)] \cdot P x a\}$

Hintikka's semantics also allows us to treat cases of the *iteration* of epistemic operators, by means of the relation of *alternativeness*: "if we are considering what *b* believes (or fails to believe) that *d* knows or does not know, we are *ipso facto* considering epistemic *d*-alternatives to these doxastic *b*-alternatives, and the requisite condition has the form

$(Ex) [(a = x) \cdot (b \text{ believes that } (a = x)]$
$[b \text{ believes that } d \text{ knows that } (a = x)]$."

Føllesdal says quite rightly in *Referential Opacity* (1966),[34] that the replacement of the absolute conception of possibility by a relative and relational conception, as advanced by Hintikka and Kanger, constitutes the chief innovation in this domain since Leibniz.

Without concern for the stand we took in favor of extensionalism, and disclaiming all doctrinal nominalism, we opted for Hintikka's solution, a choice motivated by technical considerations. It is now important to see if this theory, preferable on its own grounds, can be reconciled with positions we have previously adopted; for the use this theory makes of the notion of 'possible world' or 'possible individual' is a ground for suspicion. This notion indeed does not appear to be well-founded, neither from the *ontological* nor from the *epistemological* point of view. We shall now examine these two objections.

As individuals who people possible worlds are not accessible to observation and can only be apprehended mediately through language, one could ask whether Hintikka's *possible individuals* differ significantly from Frege's *individual concepts* which are the designata of referring expressions occurring in oblique contexts, and if the *fragments of possible worlds* differ significantly from Frege's propositions.

A careful examination reveals a fundamental difference between these two kinds of entities. A possible individual does not differ *categorially* from actual individuals, *the actual world is a subset of the union of all possible worlds. A possible individual which is realized does not cease being possible*. On the contrary, however, a non-exemplified individual concept like the concept 'Pegasus' would if exemplified remain distinct from its exemplification. Frege's concept is indeed conceived as a separate and intermediary entity between a name and the thing named. The same applies to propositions whose character as an *intermediary entity* belonging to a third world (that is, neither to the physical universe nor to the universe of signs) was denounced by Wittgenstein in the *Philosophical Investigations* (1953), when he cites a "tendency to assume a pure intermediary between the propositional *signs* and the facts".[35] Nothing of that sort can be found in Hintikka's treatment. By an individual he understands an individual member of a possible world. This can be an actual individual associated with other actual individuals in a possible combination, much like *actual* enemy batallions whose commander in chief considers *possible* deployments.

We concede that an extensional translation of 'it is possible that he come' as 'he will come in a possible world' has the appearance of an unfaithful translation. But this is only a misleading appearance. To get rid of it, one has only to think that the actual world is one among the possible worlds.

The above mentioned translation could also be faulted on the ground that it does not succeed in eliminating the world 'possible'. This objection is, however, unfair, and completely overlooks the fact that, as Hintikka puts it in 'Grammar and Logic' (1973), "by stepping from a world to its alternatives, we can reduce the truth-conditions of modal statements to truth-conditions of nonmodal statements".[36]

The notion of a possible world is moreover not an obscure notion apprehended only by such analogies as those borrowed from Hintikka that we have just used. Quine gave a precise definition in 'Propositional Objects' (1967).

> What then is a possible world? To simplify matters let us accept for a while an old-fashioned physics according to which, as Democritus held, all atoms are homogeneous in substance and differ only in size, shape, position, and motion. Let us suppose further that space is Euclidean.
>
> Now when this much is granted, there remain for each point in space just two possible states: the point may be within some particle or it may be empty. Each distribution of these states over all the points of space may be seen, not yet quite as a possible world, but as a possible momentary world state.[37]

Quine has thus arrived at a definition of the notion of a 'possible world state'. From this he constructs the notion of a 'possible world'. As the points in space are in a one-to-one correspondance with the real numbers, Quine defines a possible world as a class of classes of quadruples of real numbers corresponding to a given class of real numbers.

To avoid difficulties arising from the identification of individuals through possible worlds, Quine imposes constraints upon the latter: he works with "centered states of affairs".[38] As a consequence he can account for egocentric propositional attitudes only.

We have just seen that the notions of possible world and possible individual are *ontologically* well-founded notions, and in each case much more satisfying from this point of view than intensional entities such as objective concepts or propositions. But the notion of a possible individual has also been criticized from an *epistemological* point of view. Quine has repeatedly attacked it for not having a criterion of individuation. He considers the notion of an unactualized concrete object as still more unacceptable than intensional notions.

Hintikka has recently dissipated the uneasiness that the notion of a possible individual might engender. In *Model for Modalities* (1969), and 'The Semantics of Modal Notions and the Indeterminacy of Ontology' (1970), he refuses from the outset to postulate a universe of prefabricated possible individuals whose identity criterion would indeed be arbitrary. He substitutes for the epistemologically suspect notion of a possible individual the much more satisfying notion of an 'individuating function'. By this he understands: "a function which picks out from several possible worlds a member of their domain as the 'embodiment' of that individual in this possible world or perhaps as the *role* which that individual plays under a given course of events".³⁹

This notion of an individuating function, which we again find in Montague's pragmatics, has a definite empirical content. It presupposes that continuity properties and similarity relations obtain among the objects. Now, if these properties and relations upon which the individuating functions depend are not guaranteed by logic, then *they are guaranteed by the laws of nature*. They share with them their security and their relative precariousness.

This appeal to the laws of nature in defining the notion of a possible world is very important. It salvages a semantics which utilizes this notion from certain criticisms. If one wishes to consider, for example, the class of possible worlds which are compatible with what Albert knows, and one adds that Albert has a good background in natural science, one is free from the obligation to take into consideration fruitless logical possibilities such as a possible world in which the Moon is made of green cheese.

It would also be fitting to suppose that our subject has a degree of historical knowledge in order to discard the physically possible worlds which have ceased being possible, such as the world in which Caesar was not killed by Brutus.

10. THE PRAGMATIC APPROACH TO THE PROBLEM OF INTENSIONAL CONTEXTS: NATURAL PRAGMATICS

We noted above that the transition from indirect to direct discourse entails an alteration of information content. This can be easily verified, as Church has pointed out, by means of the translation test.

This transition form indirect to direct discourse also involves an alteration which has not often been acknowledged by logicians. An example will easily bring it out.

Consider the following two sentences:

(1)   Peter says that the president of the Republic was a tyrant.
(2)   Peter says: "The president of the Republic was a tyrant".

When making statement (1), one attributes to Peter not only a verbal pronouncement, but also an affirmation for which he is held responsible.

However, when making statement (2), one attributes to him only the formulation of a sentence and does not make it clear whether Peter gave his consent to the content of the sentence or whether he, for example, pronounced it to illustrate a grammatical point.

The difference between direct and indirect discourse comes out again in a dimension of the meaning of speech acts, termed their 'illocutionary' dimension by Austin and analysed in our Chapter IV, Section 3. In the definition of illocutionary act one takes not only the sign or its content into consideration, but the speaker as well. The theory of illocutionary acts, therefore, belongs to pragmatics in Morris' sense. Thus, it is not surprising that one has tried to resolve the logical problems of indirect discourse by considering them from the point of view of pragmatics. This track has been taken by analytic philosophers as well as formal logicians. As they both confer a great importance, though a different sense, to the concept of a proposition, it is essential that we consider their results.

In *On Referring* (1967), Linsky associates himself with the analytic school. According to him, the logician who is surprised to find in indirect discourse exceptions to the principle of substitutivity actually raises a false problem. If logicians had been attentive to the role of indirect discourse, if they had not neglected the role of hearer and speaker, that is, the pragmatic dimension of discourse, then the argument involving indirect discourse would not have taken, in their eyes, the appearance of an erratic impediment to which the usual rules of inference do not apply.

The fact that indirect discourse is refractory to the principle of substitutivity of co-referential terms

...is not immediately open to the logician's view because of his tendency to talk of 'statements' or 'propositions' in abstraction from actual speech situations which involve, along with these statements, both speakers and audiences.[40]

It is remarkable that Linsky, contrary to Frege, makes the proposition the *cause* of the illness rather than its *remedy*:

He [the logician] notes that, e.g., though it is true that Oedipus wanted to marry Jocasta and that Jocasta was the mother of Oedipus, it is not true that Oedipus wanted to marry his mother. But note that if one is speaking to an audience that knows the story there is no objection at all to saying 'Oedipus wanted to marry his mother'. One may say this last to an audience with knows that Oedipus was unaware of the fact that his mother satisfies the open sentence 'x = Jocasta'. Such an audience will not regard the statement 'Oedipus wanted to marry his mother' as false, as the logicians say it is, much less will they regard it as misleading or an unfair or inaccurate account of what Oedipus wanted.[41]

Instead of constructing a special logic for indirect discourse, Linsky proposes preserving ordinary logical syntax, governing its application by a guiding principle involving a pragmatic consideration, a principle which is stated thus: "One must never induce one's audience to error". The application of this principle clearly requires factual knowledge; we must know what our audience knows or believes.

In 'Reference and Belief' (1969),[42] Zemach also treats arguments whose sentences involve indirect discourse as enthymema in which the unstated premises contain pragmatic information relating to the language spoken by the author of the argument.

Consider the following argument:

(1)   George is the richest man in town
(2)   Jones does not believe that George is the richest man in town

Therefore:

(3)   Jones does not believe that *the richest man in town* is *the richest man in town*

Therefore:

(4)   Jones does not believe that *George* is *George*.

According to Zemach, conclusions (3) and (4) are perfectly acceptable. They seem to be paradoxical, but this is only an appearance and this appearance is due to an *equivocal* use of the definite description 'the richest man in town' or of the proper name 'George'. These terms, according to Zemach, do not have the same sense in the premises as in the conclusion. In the conclusion, the first occurrence of a designating term in the subordinate clause is to be accompanied by some indication that it belongs to *our* language or at least that it applies our criteria of reference to Jones. In the premise, on the other hand, the same designating term in this instance belongs to Jones' language or at least applies to his criteria of reference. An appropriate indication ought to mark it.

In 'Phrase et Acte de parole' (1970), Strawson offers just about the same diagnosis. Like Zemach, he maintains that in order to dissipate the 'referential opacity' in sentences of indirect discourse, or in sentences related to these, it is necessary to furnish information relating to the audience and to the speaker's intentions.

Thus, for example, the sentence 'Smith hopes that your brother will be elected' admits several interpretations. The description 'your brother' can be taken in a sense that does not imply that Smith knew that the person whose election he is hoping for is the hearer's brother (Smith hopes that the person who, in fact, is your brother will be elected). But it can also be taken in the sense that Smith knows that the person concerned is the hearer's brother. Finally, the same sentence can contain the *suggestion* that Smith only hopes for the election of the person in question *inasmuch* as that person is the hearer's brother.

In Strawson's view, sentences taken only with regard to their sense and reference, but without regard to these refinements relating to the pragmatic dimension of meaning, are not *ambiguous* but, rather "intrinsically *indefinite* in sense".[43] The additional information required to make *determinate* a sentence afflicted by referential opacity does not always reveal its *linguistic* meaning. Thus, for example, the *suggestions* are part of the complete meaning of a statement, but these vary with the situation and defy any codification.

Strawson, however, could not accept Zemach's conclusions, who relativizes the semantics of the proper name 'George' and the definite description 'the richest man in town' to Jones and the speaker.

Even if these two persons have recourse to different criteria for identifying George, even if Jones has not had the opportunity to state that the individual whom he rightly calls 'George' is spatio-temporally contiguous with the richest man in town, it does not follow, however, that their respective uses of 'George' belong to different languages, which are in some sense private — all of which Zemach would be constrained to admit.

The authors we mentioned all impute explicitly or implicitly the difficulties encountered by the logician in dealing with indirect discourse and related contexts to his abstracting linguistic material from the pragmatic dimension of language. The remedy they commend is to take the utterance as the basic unit of analysis rather than the proposition, which amounts to relativizing the semantics of a sentence to information, linguistic or otherwise, possessed by speakers and hearers.

What is the merit of this solution to the problem of the failure of the principle of substitutivity in intensional contexts? It does not appear to satisfy the demands of good theoretical explanation as we formulated them at the beginning of this study. Indeed, it explains only *one* of the peculiarities of intensional logic, namely the impossibility of substituting coreferential terms *salva veritate*.

As Genova notes in 'Jonesese and Substitutivity' (1971), Zemach's explanation leaves untouched the problem concerning the legitimacy of existential generalization.

For example, there are problems concerning the permissibility of existential generalization in respect to substantival expressions occurring in the embedded sentences exhibited in the oblique containing sentences. Does 'Jones believes that Mary is a virtuous woman' entail 'Mary exists'?[44]

Zemach's solution cannot even explain all the shortcomings involving the principle of substitutivity, notably those occasioned by modal contexts, such as 'It is possible that Venus is not the morning star'.

If the concern for generality does not seem to have much concerned the philosophers of natural language who have sketched a solution to the problem of intensional contexts within the framework of pragmatics, it is, however, a dominant concern of Richard Montague, one of the founders of formal pragmatics.

Furthermore, Montague makes use of the notion of proposition conceived as a non-linguistic entity. Does Montague's very promising research go against the stream of nominalism to which we have subscribed? An attempt to answer this question will conclude our critical examination of the various positions taken in the matter of the logic of intensional contexts.

11. THE PRAGMATIC APPROACH TO THE PROBLEM OF INTENSIONAL CONTEXTS: FORMAL PRAGMATICS

The opposition between ordinary language philosophers, who considered formalization as a distortion, and formal logicians, who scorned at ordinary language which they found too loose for the purposes of rigorous reasoning was superseded by Richard Montague who has shown it possible to encompass the syntax and semantics of both kinds of languages within a single natural and mathematically precise theory.

A detailed study of Montague theory of meaning falls outside the scope of the present inquiry. I shall restrict myself to a succinct examination of some of Montague's ideas which contribute to the solution of problems raised by itensional contexts and indirect discourse.

Before undertaking to set up a rigorous syntax and semantics of fragments of natural languages treated in the same way as formal languages, Montague made a significant contribution to the theory of meaning confined to formal languages. Quine claimed that semantics should be divided into two fields — a theory of reference and a theory of meaning — and he deplored that the second one has not given rise to a rigorous theory such as Tarskian semantics. In *Pragmatics* (1968), Montague takes up the challenge and claims that the theory of sense is amenable to a rigorous treatment within the confines of a formal pragmatics. He rejects Quine's suggestion about the division of semantics: "This suggestion", Montague writes "turns out upon investigation probably not to represent the best division: the theory of meaning, it appears, can most naturally be accommodated within pragmatics".[45]

Pragmatics as Montague conceived it in 1968 is an extended predicate calculus incremented by modal and tense operators and above all by personal pronouns and demonstratives.

The main innovation, however lies elsewhere. It lies in the interpretation. In interpreting a pragmatic language, Montague says, we have to supply four types of information: Only the first two will be considered here.

(1) we have to specify the set of all complexes of relevant aspects of intended possible contexts of use, complexes which Montague calls, after Dana Scott, *points of reference*.

(2) Another kind of information needed is the intended *intension* (or meaning).

Montague writes: "To do this for a predicate P we should determine, for each point of reference i, the *extension* of P with respect to i".[46]

To obtain this result, Montague adopts a strategy developed by Kripke in *Semantical Considerations of Modal Logic* (1963). He construes the intension of predicates as functions from indices (or possible worlds) to sets of objects and propositions as functions from indices — more specifically from possible worlds — to the set $\{0, 1\}$ of truth-values.

Pragmatics was extended by Montague in another essay, *Pragmatics and Intensional Logic* (1970), which offers new answers to several of the problems discussed in this chapter. Several differences between the two essays deserve mention. *Pragmatics and Intensional Logic* deals with second-order logic and contains pragmatics as a proper part. Pragmatics, on the other hand, can be regarded as a first-order reduction of part of intensional Logic.

Whereas *Pragmatics* expressed intensional entities, *Pragmatics and Intensional Logic* contains expressions which *denote* them.

Montague introduces descriptives phrases such as $TG\phi$ (where G is a predicate variable and ϕ a formula of the language L) which he abbreviates by the intensionalized expression. The above mentioned expression can be read 'the proposition that ϕ'. Such descriptive phrases can always be eliminated but at the cost of quantifying over predicate-variables (here the O-place predicate-variable 'ϕ' or propositional variable).

By Quinean standards, this essay is more ontologically committed than the former but this concession to Platonism is fully justified in Montague's eyes by the extra power it gives to the semanticist anxious to solve vexing problems in the logic of indirect discourse:

The proposal to regard belief as an empirical relation between individuals and propositions is not new. A number of difficulties connected with that proposal are, however, dispelled by considering it within the present framework; in particular, there remains no problem either of quantifying into belief contexts or of iteration of belief.[47]

Let us consider the following sentence which illustrates both iteration of belief and quantification into belief contexts

> There exists an object of which Jones believes that Robinson believes that it is perfectly spherical.

It will be represented by

$$Vx\,(E\,[x]\,\&\,\mathcal{B}\,[J,\,\hat{}\,\mathcal{B}\,[R,\,\hat{}\,S\,[x]]]).$$

If we eliminate the descriptive phrases which lies behind the intensifier $\hat{}$, we obtain the following sentence of second order logic:

$$VxVG\,(E\,[x]\,\&\,\mathcal{B}\,[J,\,G]\,\&\,\square\,(G\,[\,]\,\leftrightarrow\\ \leftrightarrow VH\,(\mathcal{B}\,[R,\,H]\,\&\,\square\,(H\,[\,]\,\leftrightarrow S\,[x])))) $$

where 'Vx' is the existential quantifier binding a variable ranging over the set of possible individuals, 'E' is the predicate constant 'Exists', 'B' is the dyadic predicate 'Believes', 'J' and 'R' are individual constants designating Jones and Robinson, respectively, and S is a predicate constant regarded as expressing the property of being perfectly spherical.

It should be noticed that, complicated as it looks, Montague's treatment is considerably *simpler* than Frege's treatment. According to Frege indirect speech shifts the meaning. The intension in direct speech becomes the reference in indirect speech. If we apply this principle to a belief context which itself lies within belief context, we need *indirect* intension as reference for the denoting expression (name or variable) occurring behind two Belief-predicates. In other words, the argument of 'Believes...' will not belong to the same logical level as the argument of 'x Believes that y Believes that ϕ'. This policy creates artificial differences between the domain of variables and leads to a *multiplicatio entiarum*. Montague escapes these unpalatable consequences of Frege's theory because all his propositional variables ('G' and 'H') in the formula previously mentioned range over the *same* domain: namely, over propositions. Montague's rejection of indirect intensions can thus be seen as an *application of Occam's razor*. On the other hand Montague's intensions can be viewed as set-theoretic constructs which are compatible, if not with a Goodmanian nominalism, at least with a Quinean extensional Platonism.

12. OBJECTIONS AGAINST MONTAGUE'S SEMANTICS

In spite of the economy, I have just pointed out that Montague's semantics contains an ingredient unacceptable to the nominalist: namely the notion of a class of *possible worlds*. The latter, in fact, play a crucial role in the definition of intensions since the intensions of the sentences, i.e. the propositions are conceived as functions from possible worlds to truth values. Montague's notion of possible worlds is not constrained like Quine's.

To appease the objections of the moderate nominalist (i.e. of the 'Extensionalist'), it might be argued that the situation is no worse here than in real number theory. As Barbara Partee observes in 'Semantics-Mathematics or Psychology' (1978),

> you don't need to represent all of the possible worlds distinctly in order to know a function which has them as domain. We know the function for adding arbitrary real number without being able to represent all the real numbers distinctly.[48]

One might also try to limit the set of possible worlds available to the semanticist. This has been suggested by Asa Kasher in 'Logical Rationalism. On degrees of adequacy for semantics of natural language' (1976). Psychological considerations in line with Chomsky's attempt to characterize the class of possible human languages led Kasher to introduce in this context the notion of a highly structured logical space. With the help of this notion, one can formulate a constraint limiting the class of possible worlds to which the semanticist is entitled: "The classes of possible worlds used in a semantic theory should be appropriately representable within the logical space suggested by the theory"[49] and he adds, "For intensional logic to constitue an essential part of an elementary adequate semantic theory, its framework should be finitely and uniformly representable".

The notion of proposition conceived as a function from possible worlds to truth-values together with the principle of intensionality (i.e. the principle which allows the exchange of logical equivalents) is responsible for a more serious difficulty than the objections just considered: in fact, it leads to incorrect predictions. This is easy to illustrate. If ϕ and Ψ are logically equivalent sentences, then according to Montague, the following sentence is logically true

$$B[J, \char`\^\phi] \rightarrow B[J, \char`\^\Psi]$$

where 'B' stands for 'believes', 'J' for 'John' and '$\char`\^\phi$' for 'the intension of the sentence ϕ'. But this is counterintuitive. It allows us to infer from

(1)   Irena believes that P.
(2)   P is logically equivalent to Q.

the conclusion

(3)   Irena believes that Q.

This inference is obviously invalid.

The class of counter-examples against Montague semantics constituted by the propositional attitudes is not sufficient however to condemn the whole theory. As Barbara Partee remarks:

> the difficulty of formulating an appropriate semantics for *belief*-sentences and other sentences about propositional attitudes is well-known, and I would certainly not want to suggest abandoning any semantic theory out of hand simply because that theory did not so far seem to allow any adequate treatment of the propositional attitudes.

J. Moravcsik suggests that the intensionality principle should be replaced by a stronger principle allowing the exchange of *synonymous* expressions and in some cases by an even stronger principle requiring synonymy accompanied by sameness of syntax and wording.

But in so far as it distinguishes the range of application of the two principles *in abstracto*, Moravcsik's account is still too rigid. The question of which exchangeability principle can be used cannot be answered independently of the *speaker* involved. This conclusion emerges from examples such as the one given by Tyler Burge in 'Belief and Synonymy' (1978):

> For years I believed that a
> fortnight was ten days, not fourteen,
> though of course I never believed
> that fourteen days were ten days.[50]

Such counter-examples to the principle of allowing the interchange of synonymous expression exemplify a situation where "the content of the belief has not been fully mastered by the believer" and where "the relevant words in the believer's repertoire are clearly not sufficient to determine that content apart from construal of the words".[51]

This led Barbara Partee to represent the statement of propositional attitude

Charles believes that Hesperus is Phosphorus

by the following formula

$$(\exists i)(\exists m) \text{Believes}(\text{Charles}, m(i(\text{Hesperus}) = i(\text{Phosphorus})))$$

where *m* designates Charles' own psychological performance factors and *i* his idiosyncratic interpretation of lexical items.

In the same spirit, Jennings and Schotch[52] offer independent arguments for the view in which "*de dicto* beliefs are regarded as beliefs about the meaning and reference of the terms of a sentence".

This relativization of meaning to the speaker or to the language takes us a long way from Bolzano's *Satz an sich*. It is worth emphasising that these important concessions to one of the aspects of Nominalism were motivated not by a methaphysical bias but emerged from a very methodical consideration of empirical data. They are all the more significant.

REFERENCES

[1] J. Hintikka, *Knowledge and Belief*, Cornell University Press, 1962.
[2] G. Frege, 'On Sense and Nominatum', in H. Feigl and V. Sellars (eds.), *Philosophical Analysis*, p. 93.
[3] G. Bachelard, *Le Nouvel Esprit Scientifique*, Alcan, Paris, 1939, p. 157.
[4] B. Mates, 'Synonymity', *Semantics and Philosophy of Language*, ed. by Linsky, The University of Illinois, Urbana, 1952, p. 125.
[5] R. Carnap, 'On Belief Sentences: Reply to Alonzo Church', Supplement C to *Meaning and Necessity*, 1956, p. 231.
[6] A. Church, 'On Carnap's Analysis of Statements of Assertion and Belief', *Analysis* (1950) 97–99. Reprinted in M. Macdonald (ed.), *Philosophy and Analysis*. Blackwell, Oxford, 1954.
[7] I. Scheffler, 'An Inscriptional Approach to Belief Sentences', *Analysis* (1953–1954) 87.
[8] W. V. O. Quine, *Word and Object*, p. 215.
[9] A. N. Prior, *Objects of Thought*, in P. T. Geach and A. J. P. Kenny (eds.), Oxford Clarendon Press, 1971, p. 19.
[10] Prior, *Ibid.*, p. 17.
[11] A. N. Prior, 'Oratio Obliqua', *Proceedings of the Aristotelian Society,* Supplementary Volume, 1963, p. 118.
[12] A. N. Prior *Objects of Thought*, p. 19.
[13] A. N. Prior 'Is the Concept of Referential Opacity Really Necessary?', *Acta Philosophica Fennica* (1962) 190.
[14] L. J. Cohen, 'Critical Notice: Objects of Thought', *Mind* (1973) 135.
[15] L. J. Cohen, *The Diversity of Meaning*, Methuen, London, 1962, p. 202.
[16] T. M. Simpson, 'On a nominalistic analysis of non-extensional contexts', *Logique et Analyse* 59–60, (1972) 496.
[17] Simpson, *Ibid.*, p. 498.
[18] Simpson, *Ibid.*, p. 499.
[19] G. Frege, 'On Sense and Nominatum' in H. Feigl and W. Sellars (eds.), *Readings in Philosophical Analysis*, 1949, p. 85–102.
[20] T. M. Simpson, 'Sobre la eliminacion de los contextos oblicuos', *Critica* (1967) 21–37.

[21] A. Schilpp, *The Philosophy of Rudolph Carnap*, The Library of Living Philosophers, La Salle, 1968, pp. 889–900.
[22] W. V. O. Quine, 'Quantifiers and Propositional Attitudes' *Journal of Philosophy* (1956), reprinted in *The Ways of Paradox*, Random Press, 1966, pp. 183–194.
[23] E. Bach, 'Nouns and Noun Phrases', in *Universals in Linguistic Theory*, 1968, p. 106–107.
[24] B. Loar, 'Reference and Propositional Attitudes', *Philosophical Review* (1972) 52.
[25] D. Kaplan, 'Quantifying,' In *Synthese* 19 (1968) 178–214 reprinted in *Reference and Modality* OUP Oxford, 1971, ed. Linksy, pp. 112–144.
[26] R. Montague, 'On the Nature of Certain Philosophical Entities', rep. in R. Montague, *Formal Philosophy* 1974, R. Thomason (ed.) Yale Univ. Press, p. 165.
[27] D. Davidson, 'On Saying that', *Synthese* 19 (1968) 133–136.
[28] Kaplan, *Op. cit.*, p. 185. *Reference and Modality*. p. 119.
[29] A. N. Prior, *Time and Modality*, Clarendon Press, Oxford, 1957, p. 57.
[30] Kaplan. *Op. cit.*, p. 189–190. *Reference and Modality*. p. 124.
[31] J. Hintikka, 'Modality as referential multiplicity', *Ajatus* 20, (1957) p. 61.
[32] J. Hintikka, 'Meaning as Multiple Reference', *Akten des XIX Int. Kongress für Philosophie*, Wien 1968, Tome I. p. 340–345.
[33] J. Hintikka, 'Existential and Uniqueness Presuppositions', *Models for Modalities*, Reidel, Dordrecht, 1969, p. 133.
[34] D. Føllesdal, *Referential Opacity*, Oslo, 1966.
[35] L. Wittgenstein, *Philosophical Investigations*, Blackwell, Oxford, p. 44 §94.
[36] J. Hintikka, 'Grammar and Logic, Some Borderline Problems', in J. Hintikka, J. Moravcsik, P. Suppes (eds.), *Approaches to Natural Language*, Reidel, Dordrecht 1973, p. 198.
[37] W. V. O. Quine, 'Propositional Objects' reprinted in *Ontological Relativity and other Essays*, Columbia Univ. Press, 1969, pp. 147–148.
[38] Quine, *Op. cit*, p. 154.
[39] J. Hintikka, 'The Semantics of Modal Notions and the indeterminacy of ontology', *Synthese* (1970) 412. Reprinted in D. Davidson and G. Harman (eds.), *Semantics of Natural Language*, Reidel, Dordrecht, 1972.
[40] L. Linsky, *On Referring*, Routledge and Kegan Paul, London, 1967, p. 115.
[41] Linsky, *Ibid.*, p. 115.
[42] E. M. Zemach, 'Reference and Belief', *Analysis* (1969) 11–15.
[43] P. T. Strawson, 'Phrase et Acte de Parole', *Language* (1970) 28.
[44] A. C. Genova, 'Jonesese and substitutivity', *Analysis* (1971) 101.
[45] R. Montague, 'Pragmatics', 1968, reprinted in *Formal Philosophy*.
[46] Montague, *Op. cit.*, p. 98.
[47] R. Montague, 'Pragmatics and Intensional Logic', 1969, reprinted in *Formal Philosophy*, p. 122.
[48] B. Partee, 'Semantics-Mathematics or Psychology', forthcoming in the Proceedings of the Konstanz Conference, 1979.
[49] A. Kasher, 'Formal Semantics of Natural Languages', *Philosophica* 18 (1976) 149.
[50] T. Burge, 'Belief and Synonymy', *Journal of Philosophy* (1978) 126.
[51] Burge, *Op. cit.*, p. 138.
[52] R. E. Jennings and P. K. Schotch, '*De Re* and *De Dicto* Beliefs', *Logique et Analyse* 84 (1978) 455.

CONCLUSION

The investigation carried out in this essay has taken the form of a systematic survey of the philosophical uses of the concept of proposition. My aims were two-fold. First, I have tried to show how the different uses of the concept are interrelated. Second, I have enquired into the reasons why this concept was introduced and I have attempted to find out to what extent and in what sense it can be dispensed with.

The first field in which some philosophers have claimed that the concept of proposition was needed is logic itself. Take any logical truth of the propositional calculus, if you wish to do justice to its generality, you need propositional variables to spell it out. For instance, the law of excluded middle will be formulated in this way: 'p or not p'. But as soon as variables are used, values have to be given for them and what else than propositions could play the role of values for propositional variables?

Two devious strategies have been offered which make this argument non-conclusive. The first move consists of taking the letter 'p' not as a variable which takes both values and substituends but as a schematic letter which takes substituends only. In that view 'p' ceases to be a referring expression. It plays the same role as a blank for which sentences can be substituted. Schematic letters are open to the syntactic operation of substitution but they cannot be quantified over. The second move consists of adopting the substitutional reading of quantifiers. If we do that we can quantify over 'p' without committing ourselves to propositions since the substitutional reading of quantifiers frees us of the obligation to give values to our variables.

It has been claimed that sentences eligible as substituends for two occurrences of the same schematic letters must express the same proposition, so that, after all, we are not rid of the concept of proposition. That objection can be refuted once an extensional criterion of univocity becomes available. This is the case for any language which enjoys the metatheoretical property of completeness such as, for instance, the predicate calculus. The argument runs as follows: in an equivocal language, some sentences are provable and false; therefore, by contraposition, a language where all provable sentences are true is univocal.

This consideration leads to another way of introducing the concept of proposition. Truth and falsity, some philosophers say, are properties of propositions, not of sentences. Truth transcends language. The truths of geometry discovered by Euclid are not tied up with Ancient Greek. 'True in Greek' is a solecism. As a counter-argument, one can point out that the only concept of truth for which, up to now, a non-circular definition has been given is Tarski's semantic concept of truth according to which the predicate 'true' attaches to the *name* of a sentence (as in: 'snow is white' is true) and applies to the *sentence* named by the quoted expression. To smooth out worries about truth transcending language, it is enough to refer to the so called 'disquotation role' of the predicate 'true' which shows itself nowhere more clearly than in the T-convention ('p' is true if and only if p). To those who doubt the significance of Tarski's findings for natural language on the grounds of the difference between formalized languages (which are closed) and natural languages (which are open and perhaps inconsistent), it was pointed out that one can reduce ordinary language to a consistent fragment thereof.

Declarative sentences in ordinary language can be put to different uses: they can serve to make a statement, but also to pose a question or to issue a command. The systematic study of the latter within speech-act theory, which is an essential part of pragmatics, led some philosophers to introduce the concept of a common content shared by statements, orders and questions. What is it like? Hare introduces the term "phrastics" to describe it, Stenius speaks of "sentence radical", Searle of "proposition". Each of these conflicting answers is supported by a different set of data. A new account, which covered all of them has been propounded. I argued that performative sentences such as 'I order you to leave' are *pragmatically equivalent* but not *semantically equivalent*, to non-performative sentences such as 'Leave!' and claimed that the above-mentioned sentences can *share* the same propositional content, without having the same meaning. This is possible, indeed, because the shared proposition is only a part and not the whole of the semantic content (meaning).

With Logical Atomism, new metaphysical entities, very close to propositions, were introduced, namely 'facts': remember the famous aphorism 'The world is the totality of facts, not of things'. It is claimed that if the sentence 'snow is white' is true, there is not only whiteness and snow in the world, but also something more; namely the fact that snow is white. I argued that those who subscribe to that view fall prey to a common illusion: they take for granted that syntax is a clue to ontology. What should be retained from the aforementioned aphorism is its linguistic correlate: 'Science is a collection of sentences, not of names'. That the vehicle of science is the sentence and not the name is worth noting, but it should not be interpreted as an ontological insight. The intrinsic complexity of sentences, as opposed to words, as well as the linearity of language is not a feature imposed upon language by the reality it must represent. It is due to the limitations of the human makeup, which is finite and which has to be ready to report on what happens in an infinitely diverse world. The very conditions of learning an infinite language in a finite time require that it should be made of a finite set of words which can generate, with the assistance of a finite set of syntactic rules, an infinite set of sentences.

Facts have been invoked also in the analysis of knowledge: knowledge is a two-place relation obtaining between a knowing subject and a fact. Such an analysis, however, is clearly unacceptable. It cannot be extended to belief since in the case of false belief there is no fact to serve as the object of belief. This led some philosophers to bestow upon propositions the role of objects of belief. But it seems odd to ascribe to knowledge and to belief objects of a different ontological category, facts in one case and beliefs in the other. Such categorial diversity makes it impossible to say that knowledge is true belief and condemns as non-sensical plain statements which *prima facie* seem perfectly straightforward. For instance, it becomes impossible to say 'John knows to-day some of the things he only believed yesterday' since a formal rendering of that sentence would be 'For some p; John knows p today and he only believes p yesterday' in which the variable 'p' would range over both facts and propositions and thus embody a clear case of equivocation. Some philosophers suggested that we should ascribe the same kind of object to the two propositional attitudes of belief and knowledge, namely propositions.

How can we reconcile this with the view that the bearers of truth-values are sentences? A way out would be to take sentences both as objects of propositional attitudes and bearers of truth-values. A unification of some sort is needed in any case if we want to make sense of sentences such as 'An omniscient being would know all truths' which can be formally written in this way 'For all p, if p is true, and x is omniscient, x knows p'.

The first solution one thinks of consists of giving *sentences* as value to the variable 'p' and *names of sentences* are substituends. On this analysis the sentence 'If it is the case that $\sqrt{2} = \sqrt{2}$ and if God is omniscient then God knows that $\sqrt{2} = \sqrt{2}$' would be 'If '$\sqrt{2} = \sqrt{2}$' is true and God is omniscient then God knows '$\sqrt{2} = \sqrt{2}$''. The quoted sentence which follows the verb 'knows' should create no problem. In natural language one construes 'knows' with 'that' (x knows *that* so and so is the case) and one might treat 'that' as a nominalizing device which plays the same role as the quotation marks.

There are, however, two more serious objections which stand in the way of those who take the objects of propositional attitudes to be sentences. First, one might object that mute animals such as dogs or cats are capable of entertaining propositional attitudes. Second, the supply of sentences is not rich enough to deal with the previous example: the class of true sentences is denumerably infinite whereas the class of true propositions is indenumerably infinite. Witness the set of true identity propositions about real numbers, the set of which is provably non-denumerable. One should therefore refrain from identifying those propositions which are the object (or objectives) of propositional attitudes with sentences. Hence, another candidate which could be both the object of a propositional attitude and the bearer of truth-values was sought. Beliefs understood as intentional contents took over the tasks which sentences could not carry out properly.

These intentional contents differ from Platonistic entities and from linguistic entities as well. Yet, they share many features with sentences: they can stand in logical relations like implication, mutual exclusion and incompatibility. There is even a more important and striking similarity. Belief-contents are related to reality by the same sort of relation as declarative sentences: they may represent or misrepresent reality. On this view, which I have borrowed from Searle, propositional attitudes are not directed at queer objects like Meinongian *possibilia*, but they are related to the world. What is queer if I may say so, is not the *object* or the *relatum* of propositional attitudes, but the *relation* itself, i.e. the relation which ties the intentional state to the world. This relation I called the representation relation, endorsing again Searle's view which I supported by arguments of my own.

Propositions have been invoked as *significata* of sentences on a par with individuals as *denotata* of referring expressions. Here again I argued that it is not the *relatum* of the relation of meaning which is 'queer' but the relation itself. Just as the believer tries through his belief to *represent* to himself a fragment of reality, in the same manner, the speaker through his declarative sentences tries to *represent* a fragment of reality.

If the significata of sentence are fragments of reality, how can a sentence be false and still have a meaning?

I argued that a solution to another problem, i.e. to the problem of accounting for the creativity of language, i.e. for the fact that a linguistically competent subject is able to understand sentences he has never heard before, contains part of the answer to the problem raised by the meaningfulness of false sentences. I claimed that, in both cases, the solution lies in the recognition of the combinatorial nature of the language and more especially in the compositionality principle, according to which the meaning of the whole is a function of the meaning of the part and of the syntactic structure. The idea is that once one has extracted the syntactic rules of the language out of the paradigm sentences, one is able to create new compounds which go far beyond the states of affairs for the existence of which we have evidence and a field of application opens up for guesses and risky assertions.

I have just said that the meaningfulness of false sentences was a negative effect of the productivity of language. It shows what could be called an *excess* of power on the part of sentences which enable us to describe not only the new and the possible but also the false.

The claim that the *significata* of the sentences are *fragments of reality* can nevertheless be maintained but it requires a qualification. The claim holds *only* for a particular class of sentences-tokens, namely for the paradigms which serve to teach the language. Once the language has been taught, sentences which are both false and meaningful can be framed. Yet falsity is parasitic upon truth in the way meaningfulness is parasitic upon truth.

A pragmatic explanation is needed to show that the false is parasitic upon the truth in a more important sense than the possible. Let me recall my account. Elementary sentences are endowed with meaning in paradigm cases in which they are true and *seen* to be true. The necessities of life compel us to make guesses, i.e. to assent or dissent with sentences which we do not see to be true. Suppose they are false. In that case we *misuse* the copula, i.e. we use it in a way which is not in conformity with the way in which it was used in the paradigm case. But it is a kind of misuse which is built into the language-game of stating. It is completely different from the sort of misuse illustrated by those who misinterpret an affirmative nodding as meaning a negative. The copula is liable to two quite different misuses: linguistic misuses and factual misuses; the other signs are open to the former only. Meaning, in my view, depends on truth. False sentences are parasitic upon true sentences, i.e. upon the sentences used as paradigms to teach the language. In a parallel manner, I would say that belief is parasitic upon knowledge. For there being meaning anywhere, there must be truth somewhere, for there being belief anywhere, there must be knowledge somewhere.

I have just compared the way our *propositional attitudes* (beliefs or knowledge) represent, rightly or wrongly, what is out there to the way *sentences* represent what is out there. Since resemblance is a symmetric relation, the comparison must work both ways. We have seen that the comparison of beliefs with sentences illuminates the representative character of the former. One can expect that a similar comparison will shed light on features of the latter. A well known fact about beliefs is this: there are shared beliefs. Can we say in the same way that there are shared sentences or, better, shared sentential meanings? This is a basic assumption which underlies the very task of translation: there are invariants of meaning which it is the business of the translator to transfer from one language to another. This assumption provides a new way of giving access to the concept of proposition. Just as *direction* can be obtained by abstraction from a set of lines parallel to a given line, so *proposition* can be obtained by abstraction from a set of sentences interlinguistically synonymous with a given sentence.

A necessary condition for the formulation of a definition of that kind is that we have at our disposal an equivalence relation. Is the synonymy relation symmetric, reflexive and transitive? The famous and controversial thesis of the indeterminacy of translation stands in radical opposition to the claim that there are any objective relations of synonymy. For the proponent of the indeterminacy thesis, meaning depends on empirical data about linguistic behavior but also, and irremediably, upon the manual of translation we choose. Admittedly we have criteria which guide our choice, such as the criterion of simplicity, but the latter is not a way of retrieving objectivity as it is in natural sciences, for the relation 'simpler than' defined on the set of manuals of translation is not an ordering relation. I agreed with that claim and, to the same extent, rejected the concept of proposition-as-meaning-invariant which is built upon the concept of objective synonymy.

Identity criteria for meaning are lacking and, if we take seriously the parallel drawn above between beliefs and meanings, identity criteria for beliefs should also be lacking. This expectation is fulfilled as can be seen by the breakdown of substitutivity within belief contexts. I may believe that John will come in fourteen days and yet not believe that he will come in a fortnight in spite of the fact that 'fourteen days' is synonymous with 'a fortnight'. Beliefs are somehow shaped by sentences, and may suffer from an idiosyncratic lack of synonymy of sentences in the idiolect of the believer. Other factors also account for the breakdown of the substitutivity principle, for instance, the gap between the rule of substitutivity and its effective application by the human mind, which sometimes breaks the rule, i.e. the gap between competence and performance.

This concern leads us to the last way of introducing propositions which was considered in this essay. I surveyed the recent attempts made to deal with the logic of indirect discourse. A logic designed to account for quantification into belief sentences or iteration of belief cannot do without using bound propositional variables, which means that propositions have to be assumed after all if we want to cope with the complex problem raised by the logic of belief. I stressed, however, that in spite of this commitment to propositional entities, the recent intensional logics offered were more parsimonious in ontological commitment than those previously offered, which were built along Frege's lines.

I have tried to minimize the departure from nominalism, or at least extensionalism, which the adoption of such an intensional logic entails by pointing out that if we take propositions to be functions from possible worlds to truth values, what we are left with are nothing else than set theoretic entities. Admittedly, an orthodox extensionalist will be reluctant to accept a set of possible worlds. One can, however, weaken his opposition by imposing suitable constraints upon the eligible possible worlds. One should also keep in mind that propositions, in the view presented here, are nothing else than theoretical constructions which we assume to the extent to which they are needed by a logic strong enough to account for the intuitions of validity of the speakers of natural language.

To what extent can I claim that my survey is a nominalist theory of propositions? I admitted from the start that I did not use 'theory' in the strict sense of a hypothetico-deductive system but in the loose sense of a 'coherent view' about the topic chosen. Aiming at that not very ambitious goal I asked myself how propositions should be conceived in order to fulfill the multifarious roles they were introduced to play: what should propositions be in order to do the 'premissory job', to enter into logical relations of implication and incompatibility, to be predicated of truth or falsity, to be the common content of several speech acts, to be the objects of propositional attitudes, to be the meanings of sentences and finally to be the content which remains invariant through translation.

Another form of theoretical unification should be mentioned here. In the analyses of facts, beliefs and meaning for which I argue, a positive role has been given to syntactic feature and withdrawn *to the same extent*, from semantic, psychological or ontological entities. Some sort of law of *compensation* comes into play here which might be called *syntactic ascent*.

The *unification* aimed at was realized within certain limits. I have adopted the view propounded by others to the effect that beliefs are the best candidates one can find for these various roles. How far can this view be labelled nominalist? It is nominalist to the extent to which beliefs are likened to sentences in so far as they have the capacity of representing and in so far as most of them are shaped by sentences. It is also nominalistic in so far as Bolzano's *Satz an sich* as postulated by translators was dispensed with. The unification attempted is, however, partial and by the same token the nominalistic reduction is partial also, since I was led to subscribe to propositional entities playing the role of values for the propositional variables of epistemic and doxastic logic. This concession to Platonism, however, was mitigated by the constraints imposed upon the set of possible worlds which enter the definition of the notion of proposition. Another breach in the nominalist program was conceded in order to encompass Keenan's and Faltz' compositional semantics in my framework.

The positive manifestation of my nominalist commitment can be seen in the permanent stress I have laid on the syntactic factors present in beliefs, sentences and propositions, coupled with the idea that, contrary to the metaphysical dogma underlying logical atomism, syntax is not a clue to ontology but the by-product of the capacity of the finite human mind to adapt itself and come to grips with an infinite world.

This nominalist *commitment* which I made at the start should not be confused, however, with a nominalist *bias*. The reader is here reminded of the distinction I drew in the Introduction between methodic and doctrinal nominalism.

A final remark is a point concerning the amount of originality I claim. Although most of the views defended in this essay were initially propounded by others, among the few theses for which I can claim reponsibility is my account of the meaningfulness of false sentences in terms of *syntactic* and *pragmatic* considerations the unification of speech act theory with formal semantics and the 'syntactic ascent' in the treatment of facts, beliefs and meaning. If consideration and discussion of this account leads to a better understanding of the various ideas involved then this book will have achieved its purpose.

NAME INDEX

Anderson, J. M. 21, 134
Anscombe, G. E. M. 138
Aristotle 5, 6, 8, 20, 45, 57, 75, 84
Austin, J. L. 5, 7, 49, 51, 55, 63, 65, 68, 71, 82, 83, 175
Ayer, A. J. 5, 6, 24–28, 76, 77, 95, 101, 104–106, 124, 125

Bach, E. 163
Bachelard, G. 77, 150
Bar-Hillel, Y. 1, 4, 49–51, 54, 55
Barcan (Marcus), R. 29, 30
Beardsley, E. L. 81
Belnap, N. 134
Benveniste, E. 61, 109
Berg, J. 31
Berkeley 31
Black, M. 79
Blake, R. R. 88
Boisse, L. 77
Bolzano 2, 61, 145, 183, 191
Boole 1, 18
Brentano, F. 87, 89, 91, 94
Brunot, F. 50

Cantor 9, 30, 54
Carnap, R. 4, 7, 9, 10, 20, 75, 76, 78, 79, 83, 84, 110, 111, 125, 134, 135, 138, 141, 144, 151–153, 159, 162, 164, 169
Cartwright 51–54
Castañeda, H. N. 3
Cheng 17
Chomsky, N. 5, 7, 110, 146, 181
Church, A. 5, 7, 10, 15, 22, 27, 75, 76, 99, 106, 119, 125, 137–139, 152, 153, 159, 168, 174
Cohen, L. J. 104, 156–158
Combes, M. 102–104
Cook Wilson 91–93

Cresswell, M. J. 127, 128

Davidson, D. 12, 57, 58, 84, 116–120, 138, 167
Democritus 173
Devaux, Ph. 1
Donnellan 165
Dopp, J. 5, 6
Ducasse, C. J. 1, 81

Eaton, R. 45–47
Encyclopedia Britannica 10
Euclid 104, 185

Faltz 119, 120
Fodor, J. 115, 116
Føllesdal, D. 138, 140, 141, 172
Frege, G. 26, 62, 64, 65, 71, 75, 76, 104, 109, 120, 125, 137, 141, 149, 158–164, 167, 169, 172, 175, 180, 191

Gale, R. 87, 92–94, 110
Gazdar, G. 68
Geach, P. T. 12, 61, 62
Genova, A. C. 177
Goodman, N. 7–9, 15, 133, 144, 153, 180
Granger, G. G. 38

Haack, R. 54, 55
Haack, S. 54, 55
Hamlyn 129
Hampshire 4
Hare 64, 65, 71, 186
Harman, G. 98
Hausser, R. 68, 69, 71, 120
Herder 104
Hintikka, J. 26, 120, 121, 149, 166, 169–173

NAME INDEX

Hobbes, T. 46, 75
Hubien, H. 4
Husserl, E. 25, 97, 98, 102, 104

Jennings, R. E. 183
Johnson, W. E. 45–47, 60, 61
Johnstone, H. W. 21

Kanger 172
Kant, I. 78
Kaplan, D. 166–169
Kasher, A. 181
Katz, J. 115, 116
Keenan, E. 119, 120
Keeton, M. T. 81
Kemeny, J. C. 49
Kneale, W. 23, 46, 48
Kripke, H. 179
Kuhn, O. 114
Küng, G. 11, 85

Lambert, K. 27
Langendoen, T. 115, 116
Langford 23
Leibniz, G. 1, 136, 159, 172
Lejewski 27
Lemmon, E. J. 50, 141
Leroy, M. 132
Lesniewski 8, 29
Lewis, D. 66, 68
Lewis, I. 23, 64
Linsky, L. 175
Loar, B. 164–166
Lyons 61, 114, 116, 121

Mahrenke, P. 83, 84
Malebranche, J. 88
Malthus 4
Marcus (Barcan) 29, 30
Martin, I. 26
Mates, B. 51, 52, 152
Meinong 102, 188
Mleziva, M. 142
Montague, R. 69, 119–121, 167, 174, 178–183
Montefiore, A. 2
Moore, G. E. 2, 89, 90, 91, 92, 93, 96

Morris, C. 105, 107, 151, 175
Moravcsik, J. 182
Morrison, J. C. 112

Naess, A. 133
Nuchelmans, G. 33

Occam 6, 51, 180

Partee, B. 117, 181, 182
Peano, G. 1
Pêcheux, M. 133
Perelman, Ch. 48
Plato 9, 20, 88, 103, 107
Plymouth 8, 10
Popper, K. R. 49
Potter, K. H. 16
Price, 129
Prior, A. N. 18, 29, 30, 131, 154–156, 168
Pythagorus 150

Quine, W. V. O. 7–9, 15, 16, 18–22, 25, 26, 29–31, 35–42, 47, 50, 52, 56, 58, 69, 73, 74, 77, 95, 114, 128, 131, 132, 138–140, 142, 145, 146, 153, 154, 162–169, 173, 178–180

Ramsey, F. P. 80, 81
Reichenbach, H. 43, 64
Resnick, M. 17
Ricoeur, P. 97, 98, 110
Robinson, R. 91
Rosenberg, J. F. 112
Ross, J. 67
Russell, B. 1–3, 5, 21, 31, 34, 39, 54, 56, 73–79, 81, 87, 108, 109, 124, 125, 132–137, 140, 154, 162
Ruytinx, J. 88
Ryle, G. 43, 61, 62, 89, 96, 97, 106–111, 125, 126, 129

Saussure, F. de 61, 108, 132
Scharle, T. 27, 159
Scheffler, I. 153, 154
Schick, K. 146
Schilpp, A. 159

Schlick, M. 126
Schotch, P. K. 183
Schröder 1
Scott, D. 179
Searle, J. 63–65, 69–72, 98, 99, 102, 110, 129, 142, 143, 186, 188
Sheffer 64
Simpson, T. L. 157, 158, 159
Stalnaker, R. 98
Stebbing, S. 79, 80
Stegmüller, W. 3, 4, 12, 30
Stenius 63, 66, 69, 186
Strawson, P. F. 1, 4, 40–42, 50, 51, 81, 82, 108, 116, 176, 177
Stroll, A. 51–54
Suppes, P. 144

Tarski, A. 4, 34, 46–49, 54, 56–58, 116, 117, 167, 186
Tillman, F. A. 81, 82

Vandamme, F. 113, 126
Vanderveken 142, 143
Vendler, Z. 12, 82, 83
Vuillemin, J. 5, 6, 15, 78

Waissmann, F. 128
Warner, R. 95
Warnock, G. J. 18, 19, 20, 21, 22, 25, 27, 30
Wells, R. 39, 106
Whitehead, A. N. 1, 21, 34, 80
Williamson, C. 28, 29
Wittgenstein, L. 6, 7, 30, 38, 39, 56, 66, 76–80, 98, 107–109, 111, 112, 150, 172

Young, I. 146

Zemach, E. M. 176–178

SUBJECT INDEX

abstract
 – entities 5, 7, 72
abstraction 5
 – concept 46
 Aristotle's theory of – 5, 6
 definition by – 43, 133
 – operator 138, 140
 by – 70
adequacy conditions 56, 57
adequatio rei et intellectu 56
aggregate 7, 8
ambiguity, ambiguous
 un – 17, 42, 47, 82, 89, 113, 115, 163, 177
analogy
 argument by – 70
 exemplified in linguistic competence 7
 of the chess game 107
 of the syllabes 167
analysans-analysandum 153
analytic
 – hypothesis of translation 145
analytically equivalent 8
analytically true 10
analyticity 135
arbitrariness 20, 135, 142
assert-assertion-assertable 62–64
assumption 15, 32
 – of context independence 121
 ontological – 16, 17, 37
 – opposed to affirmation of existence 17
 – opposed to 'presuppose' 20
atomism
 logical – 38, 56, 73, 74
 propositional – 38, 39
autonomous
 – existence of propositions 61
 – existence of statements 52
 – meaning 15, 18, 118

 – object 103
axiom, axiomatized, definition by axioms 27, 34, 35, 43, 156

bearer of truth-value 1, 55, 95, 96
behavioural
 – manifestations 152
behaviourist
 – properties 91
 – theory of meaning 105
belief, believe 2, 28, 31, 74, 83, 87–99, 83, 102, 105, 141, 149, 152, 153–183, 189
 iterated – 83, 155, 171
binarisme 132

category 5, 6, 31, 76, 84, 92, 93, 111, 112, 118, 119, 124
cause, causally, causality 35, 81, 103, 138
circularity 35
class 1, 8, 16, 75 *vide* set
coextensionality 135–138
 isomorphic – 137
coextensive 4, 9, 137, 138, 141, 159
 – predicate 9
colour spectrum 132
combination
 – of subject with predicate 84
commitment 192
 existential – 15
 illocutionary – 143
 metaphysical – 40
 ontological – 15–32, 37, 118–120
 vide countenance
committed
 – to propositions 13
compensation
 – between syntactic and semantic factors 49

SUBJECT INDEX

composition, compositional, compositionality 9, 68, 69, 116–121, 167
concept, conceptualist 53
  absolute – 46
  epistemological – 60
  – formal 102
  meta – 78
  pseudo – 78
  relative – 46
conceptualism, conceptually 53, 88
congruence
  – illocutionary 142, 143
  – in meaning 144
connective 131, 149, 155, 156, 159, 160
constants
  individual – 18
  logical – 38, 149
  predicate – 18
constituent
  ultimate – of reality 5
constraint
  – of logic 38
  syntactic – 6
context 48, 50, 74
  belief – 150–183, 190
  epistemic – 169
  – independence 121
  modal – 151, 169, 178
  oblique – 159–181
  opaque – 69, 162–181
  synonymy relativized to – 133
  sentential – 84
Convention T 118, 186
copula 84
coreferential
  – expressions 141
  interchangeability of – expressions 4
correspondence
  – between propositions facts 56, 57, 58, 75, 97, 125
countenance
  – truth-values 41
create, creative, creativity 3, 58, 109, 134, 141, 188

de dicto 152, 183
de re 152, 183
definition
  axiomatic – 34, 35
  – by abstraction 43, 133
  by genus and specific difference – 43
  – of truth 48
  operational – 131
  partial – 12
denote, denotation, denotational, denotatum 35, 36, 52, 69, 107, 109, 119, 188
  vide designatum, significatum
description 5, 81
  attributive versus referential use of definite – 165
  definite – 21, 22, 80, 140
  – operator 140
  synonymous – 80
  theory of – 21
designate, designator, designation 21, 25, 81, 82, 93, 106, 107, 154, 159
  vide denotatum, significatum
diagonal
  Cantor's – 9, 54
difference 37, 42, 75, 89, 91
  between statement and sentences 52
  categorial – 93, 112
  – of level 48
  meaning – 133, 147
  quantitative versus structural – 75
  vide independence, autonomous
dispensability
  – of signs 6
disquotation 56, 186
  vide quotation
doctrinal
  – nominalism versus methodological nominalism 5, 6, 149

economical
  – solution 41
economy
  law of – 5, 6
  vide parsimony

enthymeme 39, 171, 176
entities 32, 53
  abstract – 5, 7, 72, 153
  hybrid – 154
  independent – 2
  linguistic – 6, 71, 109
  Meinongian – 102
  metaphysical – 6
  multiplication of – 9, 180
  neutral – 125
  obscure – 50
  ontological – 6, 191
  phonetic – 102
  Platonic – 6, 154
  Platonistic – 102, 129
  – pointing to something other than themselves 96
  propositional – 74, 149
  queer – 102
  self-subsistent – 53, 61
  semantic – 102, 191
  self-sufficient – 96
  substantial – 96
  syntactic – 102
epistemology, epistemological, epistemic 60, 78, 91, 131, 133, 145, 150, 158
equiform 133
equivalence 16, 106
  analytic – 9
  extensional – 106, 160
  L- extensional – 10, 134, 135
  logical – 139
  material – 95, 138, 139
  pragmatic – 71
  strict – 143
equivalent
  analytically – 153
  pragmatically – 69, 70
  semantically – 63, 68, 69
  strictly – 155
  vide synonymous
equivocal, equivocation 29, 81, 94, 102, 161, 176
  vide ambiguity, univocal
error
  – linguistic 60, 129
  – factual 60, 129

existence
  – of an event 80
  – of facts 74
  – of numbers 54
  – statement 20
  independent – 88
  mental in – 89
  mentalistic – 94
  non-existent object 98
  the notion of – 27
  presupposition of – 27
  psychological – 89
  verbal – 61
existential
  – commitment 15
  – generalization 25, 26, 161, 164, 165
  – quantifiers 19, 30
extension 8, 35, 75, 84, 85, 158
  – opposed to intension 8
extensional 42, 99, 106, 119
  – equivalence 106
  – isomorphism 12, 138–142
  non-extensional contexts 156
  – language 159
  – logic 149
extensionalism 4, 7, 8, 150
extensionalist 4
extensionalistic
  – Platonism 120
extensionality
  – principle 18, 136, 150, 155, 156

fact 1, 2, 45, 46, 56, 57, 60, 73–85, 97, 111, 125, 127, 129, 187
fallacy
  – of subtraction 69
fiction 124, 126
'Fido'-Fido theory 107, 108, 118
field
  semantic – 16, 132
force
  causal – 81
  illocutionary – 12, 65, 143
  – of the argument 79
form
  – of the sentence 48
  grammatical – 39, 50

logical – 12, 39
surface – 71
formal
  – concept
  – logic 50
  – mode of speech 78
  – pragmatics 170–183
  – semantics 61
formalize 24, 25, 46, 155, 157, 178
  – language 27, 46
  how to – mixed inferences 171
  semi- – 24
function 38, 50
  – of language, of discursive thought 89
  propositional – 34, 58, 164
  sentential – 57
  truth – 12
  – from possible worlds into truth-values 69
  – of the proposition 76
  grammatical – 112
  meaning of a whole – of the meanings of the parts 167
functor 134

grammar
  – of natural language 119
  categorial – 119
  logical – 80
  transformational – 77, 82
grammatical
  – form 39, 50
  – function 112
  – mood 65
  – relations 114, 116, 119
  – subject 47
  – syntax 92
  – transitivity 87
  un- – 47

holism 40
homomorphism 119
homonymy 40, 75
  fortuitous – 104
hypernominalism 153
hypostatizing
  – obscure entities 50

idealism 56, 104, 116
  vide realism, pragmatism
identical
  – things, entities 52
  – morpheme 53
identification 55, 102, 131–147
identity 27, 52, 53, 85
  – for classes 8
  absolute – 133
  – criterion 8
  contingent – 140
  extensional – 41
  non-reflexive – 27
  operation with – 88
  propositional – 40, 41, 43, 44, 131–135, 145
  – of propositions 113, 156
  relation of – 33
  typographical – 40
  intensional – 135
ignoratio elenchi 92
illocutionary
  – acts 63, 64, 129
  – congruence 142, 143
  – forces 65, 70, 143
  – level 69
  – logic 142
  – point 143
imperative 48
  logic of – 1
  – sentence 2, 43, 71
  – tune of voice 65
impressionism
  linguistic – 21
independence
  content – 121
  – of sentence with respect to statements 72
independent
  fact – of language 74
  – existence 88
  logically – of the use of symbols 77
  – reasons 51
  vide autonomous
independently
  proposition exist – of thought 73
indeterminacy

– of translation 145–147
indexical
  – expressions 50, 51
individual 7, 8, 16, 18, 29, 56–58, 74, 75, 117, 119
inference 1, 43, 52, 54
  illegitimate – 150, 151, 152
  invalid – 26, 182
  mixed – 171
  modal – 151
  modus ponens – 65
  premise or conclusion of – 43
  valid – 25, 150, 163
infinite (set)
  – list 30
  – number of primitive predicates 167
  – set 30
  – space 78
  non-denumerable – 9, 30, 31, 78
intension 8, 42, 75, 85, 158, 159
  – opposed to extension 8
  quasi – 161
intensional
  – isomorphism 10, 12, 134–135, 151, 155
  – logic 4
intensionalism 7, 9
intensionality 140
  principle of – 9, 139, 150, 151
intentional
  – inexistence 89
  – objects 90
  – state 98
  – state as extensional 99
  – verbs 25
intentionality 2, 87, 89, 98, 99, 101, 102, 104
interpretation
  homogeneous – of variables 29, 31
  mixed – of variables 31
  ontological – 38
  substitutional – 29, 31, 155, 185
  vide equivocal
invariance
  – of meaning with respect to truth 124
  – of meaning with respect to translation 145

invariant
  propositional – 42, 124
  conceptual – 42
isomorphism 134–142
  extensional – 12, 138–142
  intensional – 10, 12, 134, 135, 151, 155
  principle of – 110
  – requirement 144
  requirement of – 111
  semantic – 132

judgement 1, 4, 43, 45, 46, 61, 103
  polyadic theory of – 108
  judgmental activity 103

knowledge 38, 91, 129, 189
  – by acquaintance 2
  – by description 2
  theory of – and proposition 1, 2

language
  colloquial – 48
  every day – 7
  formal – 22, 46, 47
  formalized 4, 48, 49, 54
  level of – 47, 48
  meta – 46, 49, 72, 77–81, 131
  natural – 4, 22, 23, 40, 42, 48, 49, 119, 155
  object – 46
  symbolic – 49
lexicographer 145
lexicologists 104
logic
  extensional – 149
  illocutionary – 142
  – of identity 52
  – of imperatives 1
logical
  – atomism 73, 74, 192
  – consequence 12
  – constructions 5
  – form 12, 39, 75
  – grammar 39, 80
  – multiplicity 62
  – phagocytosis 139
  – relation 98

SUBJECT INDEX

— syntax 92
— truth 4, 41, 42, 46, 48

Malthusian
 — attitude 4
meaning
 autonomous — 15, 18
 mathematical — 21
 —-postulates 10
Meinongian
 — entities 102
mereology 8
metalanguage
 vide language
metaphysical
 — commitment 40
 — considerations 70
 — entities 6, 56
 — import 75
metaphysics 84
methodological
 — considerations 183
 — dictum 132
 — exigencies 3
 — nominalism 5, 6, 77, 149
 — order 150
 — principle 30
modal
 — context 150
 — inferences 151
 — involvement 138
 — logic 179
 — operators as eliminating ontological commitment 18
mode of speech
 vide speech
mood
 grammatical — 65, 66
 Hausser's treatment of — 68, 69

name 1, 16, 28, 29, 30, 36, 52, 54, 73, 75, 85, 95, 106, 108, 109, 140, 156, 176
noema 102
neustics 64, 65
 vide phrastics, tropic

neutral
 ontologically — 27, 52
nominalism, nominalist
 doctrinal — versus methodological 5, 6, 7, 103, 149, 153, 181, 191, 192
nominalization
 — of a sentence 95
nominalized
 — sentences 23, 71
number 9, 20, 21, 30, 31, 34, 38, 54, 78, 99

object
 autonomous — 103
 degenerate — 89
 real — 94
 constructed — 5
 fictitious — 94
 grammatical — versus grammatical subjects 119
 intentional — 90
 non-existent — 98
 of belief — 90
 set-theoretic — 120
objektive 94, 95, 103
ontological
 — assumption 16
 — category 6
 — commitment *vide* commitment 13, 15–32, 36, 72, 118, 119
 — conception of facts 77
 — conclusions 6
 — controversy 2, 6
 — dichotomy 8
 — entities 6, 7
 — implication 11
 — import 15, 18, 25, 28, 147
 — inferences 6
 — interpretation 38
 — neutrality 27, 52
 — questions 20, 85
 — reality 80
 — sense 79
 — status 29, 50, 81, 87, 89
ontologically neutral
 vide neutral

SUBJECT INDEX

ontologico-grammatical
  – parallelism 5
opacity
  – of context 69, 131, 155, 163
  – of position, of zone 163
opaque
  – context 69
  vide oblique
operation
  – of exportation 165
  – of validation 39
  – with identity 88
operational
  – definition 131
  – distinction 27
operationally 43
operative
  in – 28
  – virtues of schematic letters 36
operator
  abstraction – 138, 140
  iterated epistemic – 171
  modal – 18

paradox 49
  – of the liar 49, 55
parallelism
  ontologico-grammatical – 5
parsimony
  law of – 5, 6, 31
  vide economy
performative 12, 49, 63
  – hypothesis 66, 67, 68, 169
  – sentences 67, 68
phagocytosis
  logical – 137
phrastic
  – element 65, 186
  vide tropic, neustics
picture
  – theory 109–113
Platonic
  – entities 6
  – metaphysics 9
  – philosophizing 20
Platonism 10, 120, 192
  anti- – 4

extensionalistic – 120
intensional – 10
propositional – 159
Platonistic
  – concept meaning 103
  – conception of proposition 2, 104
  – entities 102, 129, 154
Platonists 15
Plato's dialectics 9
possibilia Meinongian 188
postulate 34, 52, 108, 118, 153, 191
  meaning – 10
postulation 6
pragmatic
  – approach 151
  – component 128
  – considerations 192
  – criterion 70
  – dimension 127
  – implication 104
pragmatically 20
  – equivalent 69, 186
pragmatics 31, 60, 61, 174–183
pragmatism 20, 56
predicate 9, 15, 16, 18, 36, 45, 46, 51, 55, 60, 75, 107, 108, 119, 154, 164
presupposition, presuppose 26, 35, 38, 42, 52
primitive
  infinite number of – predicates 131
  – predicates 18, 167
  – illocutionary forces 143
  – terms 34
principle
  methodological – 30
  – of compositionality 69, 120, 121
  – of epistemological realism
  – of extensionality 8, 9, 136, 150, 151, 155, 156
  – of identification 150
  – of identity of individuals 19
  – of intensionality 9, 136, 139, 150, 151
  – of intensional isomorphism 151
  – of isomorphism 110
  – of substitutivity of identicals 139, 159, 162, 178
  – of truth-functionality 136

rational – 147
projected
  sense is – from signs to things 126
projection 115, 116, 126
  vide selection

quantification
  theory of – 16, 17, 26
quantified
  – variable 15, 18, 23, 24
quantifiers 149
  existential – 19, 23, 25, 30, 185
quantify over 24, 37, 120, 185
questions 2
  internal – versus external – 20
quotation 47, 62, 167, 187
  vide disquotation

radicals
  sentence – 63, 66, 186
real
  – number 99
realism
  – opposed to nominalism 15
  epistemological – 88
  logical – 73
  neo- – 2, 39, 51
  – -idealism controversy 116
realist 50, 51
  – account 72
  – conception of a statement 51, 53
reality 5, 6, 47, 56, 57, 124, 126, 189
  combination in – 84
  ontological – 80
  physical – 146
  objective – 132
  – as a whole 58
reasonings 26, 27, 38, 83, 90, 149
  Moore's – 90
recursive
  – account 58
  – definition of language 143
  – definition of truth 116, 117
  – definition of well formed formulae 118
  – procedure 57
  – rule 47

– semantic characterization 117
– semantics 116
– set of proposition 61
– syntax 61, 116, 119
reducing
  – universals 9
reductio ad absurdum 69
reduction
  – of propositions to sentences 152
reductionism 7
  vide absorption of propositions into sentences 96–98
regiment, regimentation 22, 142
reification 15, 121, 156
relate
  meaning – linguistic entities to the world 109
relation
  (a)symmetric – 111, 189
  binary – 89
  denotation – 108
  equivalence – 133
  grammatical – 114, 115, 116, 119
  logical – 50, 98, 119, 191
  meaning – 102, 188, 190
  name of – 111
  – of belief 90
  – of correspondence 58
  – of illocutionary congruence 143
  – of knowledge 89, 91, 101
  – of representation 98
  – of resemblance and difference 38
  semantic – 107
  spatial – 112
  temporal – 112
  translation – 145
represent, representation, representative 41, 98, 102, 110, 111, 120, 149, 189
rules
  formation – 61
  projection – 115
  – of inference 159
  semantic – 135
  truth – 156

satisfaction 57, 58, 117
Satz an Sich 145, 146, 183, 191

SUBJECT INDEX

schema
  argument – 41
schematic
  – letters 23, 24, 28, 29, 35–38, 41, 52, 185
schematism 38
schematization 36
scholastic 16, 89
selection 115
  vide projection
semantic
  – approaches 151
  Carnap's dualistic – 159–161
  – equivalence 63, 67, 68, 69, 186
  vide synonymous
  – field 16, 132
  Frege's dualistic – 158, 159
  – relation 75, 107, 108
semantics
  formal – 61
  game-theoretic – 120, 121
  model theoretic – 121
  set-theoretic – 120
sentence
  – radical 63, 66, 186
  – -type versus – -token 50, 55, 109
significatum 6, 102, 113, 129, 188
speech
  – act 98
  unit of – 61
  formal mode of – 78, 93
  material mode of – 93
structural
  – difference 75, 84
structure
  iconic – 113
  logical – 91
  semantic – 50
  – of facts similar to the – of propositions 81
  surface – as opposed to deep – 67, 71, 72, 114
structured
  – complex 113
subject
  combination of – and predicate 84
  grammatical – 47

– takes himself as – 103
thinking – 102, 103
substance 6, 79, 96, 97
substantial
  – entity 96
substitutional
  vide interpretation
substitutivity 136, 138, 177, 178, 190
  – of identicals 139
subtraction 70
  fallacy of – 69
symbol
  incomplete – 140
  propositions as mere – 73
  quasi – 126
symbolic
  – language 49
syncategoreme 15, 16, 30
syncategorematic
  – signs 18
  – treatment of opaque constructions 69
  vide categorematic – 137
synonymous 79, 82
  – descriptions 80, 182
synonymy 40, 42, 63, 71, 133, 160, 190
  vide resemblance in sense
syntactic 192
  – constraints 6
  – methods 151
  – notions 43
  – structure 96
  – transformation 163
syntax
  grammatical – 92
  logical – 38, 92–94, 152
system 35
  axiom – 39
  formalized – 48
  simplest – for correlating translations 147
systematic
  – argument 2
  – compatibility of the thing with the set of facts 150
  – organization 3
  – resolution of problem of epistemo-

SUBJECT INDEX

logic 166
- survey 185
systematically 168
  treating − the ontological aspects of the proposition 56
  − contribute to the truth-conditions 117
systematicity 12

theory 3, 104
  general − 56, 58, 76
  quantification − 17, 26
  ontology of − 16
  − opposed to rule 6
  systematic compatibility of a − with the facts 150
  unified − 50, 155
  vide 'Fido'-Fido theory and description
transcendental 77, 78, 79
transcending, transcends
  − all languages 144, 185
  − linguistic particularities 46
  − particular languages 48
  − the sentence 47
transfinite
  − sets 9
  vide infinite and numbers
transformation, transformational 5, 77, 82, 114
translation 2, 11
tree 113, 114
  amalgamation of − 115
  production − 144

tropic 65
  vide phrastics, neustics

uncountable 9
  see non-denumerable 31
understanding 125
universals
  the number of − 9
  the problem of − 3, 12, 15, 37
  the status of − 20
univocality 41, 42
univocally
  − interpreted sentences 41
  vide univocity
univocity 21, 184
Ursprache 47

valid, validity, validly, validation 16, 17, 25, 26, 39, 101, 107, 145, 157
verbs, adverbs, preposition 42
  cognitive − 92, 94
  intentional − 25, 26
  logical syntax of the − 'to believe' 152
  performative − 63
  propositional − 92, 94
  relational − 25, 26
  transitive-intransitive − 119

world 77
  actual − 127, 128
  possible − 170, 172, 173, 179, 181

SYNTHESE LIBRARY

Studies in Epistemology, Logic, Methodology,
and Philosophy of Science

Managing Editor:
JAAKKO HINTIKKA (Florida State University)

Editors:
DONALD DAVIDSON (University of Chicago)
GABRIEL NUCHELMANS (University of Leyden)
WESLEY C. SALMON (University of Arizona)

1. J. M. Bochénski, *A Precis of Mathematical Logic*. 1959.
2. P. L. Guiraud, *Problèmes et méthodes de la statistique linguistique*. 1960.
3. Hans Freudenthal (ed.), *The Concept and the Role of the Model in Mathematics and Natural and Social Sciences*. 1961.
4. Evert W. Beth, *Formal Methods. An Introduction to Symbolic Logic and the Study of Effective Operations in Arithmetic and Logic*. 1962.
5. B. H. Kazemier and D. Vuysje (eds.), *Logic and Language. Studies Dedicated to Professor Rudolf Carnap on the Occasion of His Seventieth Birthday*. 1962.
6. Marx W. Wartofsky (ed.), *Proceedings of the Boston Colloquium for the Philosophy of Science 1961-1962*. Boston Studies in the Philosophy of Science, Volume I. 1963.
7. A. A. Zinov'ev, *Philosophical Problems of Many-Valued Logic*. 1963.
8. Georges Gurvitch, *The Spectrum of Social Time*. 1964.
9. Paul Lorenzen, *Formal Logic*. 1965.
10. Robert S. Cohen and Marx W. Wartofsky (eds.), *In Honor of Philipp Frank*. Boston Studies in the Philosophy of Science, Volume II. 1965.
11. Evert W. Beth, *Mathematical Thought. An Introduction to the Philosophy of Mathematics*. 1965.
12. Evert W. Beth and Jean Piaget, *Mathematical Epistemology and Psychology*. 1966.
13. Guido Küng, *Ontology and the Logistic Analysis of Language. An Enquiry into the Contemporary Views on Universals*. 1967.
14. Robert S. Cohen and Marx W. Wartofsky (eds.), *Proceedings of the Boston Colloquium for the Philosophy of Science 1964-1966. In Memory of Norwood Russell Hanson*. Boston Studies in the Philosophy of Science, Volume III. 1967.
15. C. D. Broad, *Induction, Probability, and Causation. Selected Papers*. 1968.
16. Günther Patzig, *Aristotle's Theory of the Syllogism. A Logical-Philosophical Study of Book A of the Prior Analytics*. 1968.
17. Nicholas Rescher, *Topics in Philosophical Logic*. 1968.
18. Robert S. Cohen and Marx W. Wartofsky (eds.), *Proceedings of the Boston Colloquium for the Philosophy of Science 1966-1968*. Boston Studies in the Philosophy of Science, Volume IV. 1969.

19. Robert S. Cohen and Marx W. Wartofsky (eds.), *Proceedings of the Boston Colloquium for the Philosophy of Science 1966-1968*. Boston Studies in the Philosophy of Science, Volume V. 1969.
20. J. W. Davis, D. J. Hockney, and W. K. Wilson (eds.), *Philosophical Logic*. 1969.
21. D. Davidson and J. Hintikka (eds.), *Words and Objections. Essays on the Work of W. V. Quine*. 1969.
22. Patrick Suppes, *Studies in the Methodology and Foundations of Science. Selected Papers from 1911 to 1969*. 1969.
23. Jaakko Hintikka, *Models for Modalities. Selected Essays*. 1969.
24. Nicholas Rescher et al. (eds.), *Essays in Honor of Carl G. Hempel. A Tribute on the Occasion of His Sixty-Fifth Birthday*. 1969.
25. P. V. Tavanec (ed.), *Problems of the Logic of Scientific Knowledge*. 1969.
26. Marshall Swain (ed.), *Induction, Acceptance, and Rational Belief*. 1970.
27. Robert S. Cohen and Raymond J. Seeger (eds.), *Ernst Mach: Physicist and Philosopher*. Boston Studies in the Philosophy of Science, Volume VI. 1970.
28. Jaakko Hintikka and Patrick Suppes, *Information and Inference*. 1970.
29. Karel Lambert, *Philosophical Problems in Logic. Some Recent Developments*. 1970.
30. Rolf A. Eberle, *Nominalistic Systems*. 1970.
31. Paul Weingartner and Gerhard Zecha (eds.), *Induction, Physics, and Ethics*. 1970.
32. Evert W. Beth, *Aspects of Modern Logic*. 1970.
33. Risto Hilpinen (ed.), *Deontic Logic: Introductory and Systematic Readings*. 1971.
34. Jean-Louis Krivine, *Introduction to Axiomatic Set Theory*. 1971.
35. Joseph D. Sneed, *The Logical Sstructure of Mathematical Physics*. 1971.
36. Carl R. Kordig, *The Justification of Scientific Change*. 1971.
37. Milic Capek, *Bergson and Modern Physics*. Boston Studies in the Philosophy of Science, Volume VII. 1971.
38. Norwood Russell Hanson, *What I Do Not Believe, and Other Essays* (ed. by Stephen Toulmin and Harry Woolf). 1971.
39. Roger C. Buck and Robert S. Cohen (eds.), *PSA 1970. In Memory of Rudolf Carnap*. Boston Studies in the Philosophy of Science, Volume VIII. 1971.
40. Donald Davidson and Gilbert Harman (eds.), *Semantics of Natural Language*. 1972.
41. Yehoshua Bar-Hillel (ed.), *Pragmatics of Natural Languages*. 1971.
42. Sören Stenlund, *Combinators, λ-Terms and Proof Theory*. 1972.
43. Martin Strauss, *Modern Physics and Its Philosophy. Selected Papers in the Logic, History, and Philosophy of Science*. 1972.
44. Mario Bunge, *Method, Model and Matter*. 1973.
45. Mario Bunge, *Philosophy of Physics*. 1973.
46. A. A. Zinov'ev, *Foundations of the Logical Theory of Scientific Knowledge (Complex Logic)*. (Revised and enlarged English edition with an appendix by G. A. Smirnov, E. A. Sidorenka, A. M. Fedina, and L. A. Bobrova.) Boston Studies in the Philosophy of Science, Volume IX. 1973.
47. Ladislav Tondl, *Scientific Procedures*. Boston Studies in the Philosophy of Science, Volume X. 1973.
48. Norwood Russell Hanson, *Constellations and Conjectures* (ed. by Willard C. Humphreys, Jr.). 1973.

49. K. J. J. Hintikka, J. M. E. Moravcsik, and P. Suppes (eds.), *Approaches to Natural Language.* 1973.
50. Mario Bunge (ed.), *Exact Philosophy – Problems, Tools, and Goals.* 1973.
51. Radu J. Bogdan and Ilkka Niiniluoto (eds.), *Logic, Language, and Probability.* 1973.
52. Glenn Pearce and Patrick Maynard (eds.), *Conceptual Change.* 1973.
53. Ilkka Niiniluoto and Raimo Tuomela, *Theoretical Concepts and Hypothetico-Inductive Inference.* 1973.
54. Roland Fraissé, *Course of Mathematical Logic –* Volume 1: *Relation and Logical Formula.* 1973.
55. Adolf Grünbaum, *Philosophical Problems of Space and Time.* (Second, enlarged edition.) Boston Studies in the Philosophy of Science, Volume XII. 1973.
56. Patrick Suppes (ed.), *Space, Time, and Geometry.* 1973.
57. Hans Kelsen, *Essays in Legal and Moral Philosophy* (selected and introduced by Ota Weinberger). 1973.
58. R. J. Seeger and Robert S. Cohen (eds.), *Philosophical Foundations of Science.* Boston Studies in the Philosophy of Science, Volume XI. 1974.
59. Robert S. Cohen and Marx W. Wartofsky (eds.), *Logical and Epistemological Studies in Contemporary Physics.* Boston Studies in the Philosophy of Science, Volume XIII. 1973.
60. Robert S. Cohen and Marx W. Wartofsky (eds.), *Methodological and Historical Essays in the Natural and Social Sciences. Proceedings of the Boston Colloquium for the Philosophy of Science 1969-1972.* Boston Studies in the Philosophy of Science, Volume XIV. 1974.
61. Robert S. Cohen, J. J. Stachel, and Marx W. Wartofsky (eds.), *For Dirk Struik. Scientific, Historical and Political Essays in Honor of Dirk J. Struik.* Boston Studies in the Philosophy of Science, Volume XV. 1974.
62. Kazimierz Ajdukiewicz, *Pragmatic Logic* (transl. from the Polish by Olgierd Wojtasiewicz). 1974.
63. Sören Stenlund (ed.), *Logical Theory and Semantic Analysis. Essays Dedicated to Stig Kanger on His Fiftieth Birthday.* 1974.
64. Kenneth F. Schaffner and Robert S. Cohen (eds.), *Proceedings of the 1972 Biennial Meeting, Philosophy of Science Association.* Boston Studies in the Philosophy of Science, Volume XX. 1974.
65. Henry E. Kyburg, Jr., *The Logical Foundations of Statistical Inference.* 1974.
66. Marjorie Grene, *The Understanding of Nature. Essays in the Philosophy of Biology.* Boston Studies in the Philosophy of Science, Volume XXIII. 1974.
67. Jan M. Broekman, *Structuralism: Moscow, Prague, Paris.* 1974.
68. Norman Geschwind, *Selected Papers on Language and the Brain.* Boston Studies in the Philosophy of Science, Volume XVI. 1974.
69. Roland Fraissé, *Course of Mathematical Logic –* Volume 2: *Model Theory.* 1974.
70. Andrzej Grzegorczyk, *An Outline of Mathematical Logic. Fundamental Results and Notions Explained with All Details.* 1974.
71. Franz von Kutschera, *Philosophy of Language.* 1975.
72. Juha Manninen and Raimo Tuomela (eds.), *Essays on Explanation and Understanding. Studies in the Foundations of Humanities and Social Sciences.* 1976.

73. Jaakko Hintikka (ed.), *Rudolf Carnap, Logical Empiricist. Materials and Perspectives.* 1975.
74. Milic Capek (ed.), *The Concepts of Space and Time. Their Structure and Their Development.* Boston Studies in the Philosophy of Science, Volume XXII. 1976.
75. Jaakko Hintikka and Unto Remes, *The Method of Analysis. Its Geometrical Origin and Its General Significance.* Boston Studies in the Philosophy of Science, Volume XXV. 1974.
76. John Emery Murdoch and Edith Dudley Sylla, *The Cultural Context of Medieval Learning.* Boston Studies in the Philosophy of Science, Volume XXVI. 1975.
77. Stefan Amsterdamski, *Between Experience and Metaphysics. Philosophical Problems of the Evolution of Science.* Boston Studies in the Philosophy of Science, Volume XXXV. 1975.
78. Patrick Suppes (ed.), *Logic and Probability in Quantum Mechanics.* 1976.
79. Hermann von Helmholtz: *Epistemological Writings. The Paul Hertz/Moritz Schlick Centenary Edition of 1921 with Notes and Commentary by the Editors.* (Newly translated by Malcolm F. Lowe. Edited, with an Introduction and Bibliography, by Robert S. Cohen and Yehuda Elkana.) Boston Studies in the Philosophy of Science, Volume XXXVII. 1977.
80. Joseph Agassi, *Science in Flux.* Boston Studies in the Philosophy of Science, Volume XXVIII. 1975.
81. Sandra G. Harding (ed.), *Can Theories Be Refuted? Essays on the Duhem-Quine Thesis.* 1976.
82. Stefan Nowak, *Methodology of Sociological Research. General Problems.* 1977.
83. Jean Piaget, Jean-Blaise Grize, Alina Szeminska, and Vinh Bang, *Epistemology and Psychology of Functions.* 1977.
84. Marjorie Grene and Everett Mendelsohn (eds.), *Topics in the Philosophy of Biology.* Boston Studies in the Philosophy of Science, Volume XXVII. 1976.
85. E. Fischbein, *The Intuitive Sources of Probabilistic Thinking in Children.* 1975.
86. Ernest W. Adams, *The Logic of Conditionals. An Application of Probability to Deductive Logic.* 1975.
87. Marian Przelecki and Ryszard Wójcicki (eds.), *Twenty-Five Years of Logical Methodology in Poland.* 1977.
88. J. Topolski, *The Methodology of History.* 1976.
89. A. Kasher (ed.), *Language in Focus: Foundations, Methods and Systems. Essays Dedicated to Yehoshua Bar-Hillel.* Boston Studies in the Philosophy of Science, Volume XLIII. 1976.
90. Jaakko Hintikka, *The Intentions of Intentionality and Other New Models for Modalities.* 1975.
91. Wolfgang Stegmüller, *Collected Papers on Epistemology, Philosophy of Science and History of Philosophy.* 2 Volumes. 1977.
92. Dov M. Gabbay, *Investigations in Modal and Tense Logics with Applications to Problems in Philosophy and Linguistics.* 1976.
93. Radu J. Bogdan, *Local Induction.* 1976.
94. Stefan Nowak, *Understanding and Prediction. Essays in the Methodology of Social and Behavioral Theories.* 1976.
95. Peter Mittelstaedt, *Philosophical Problems of Modern Physics.* Boston Studies in the Philosophy of Science, Volume XVIII. 1976.

96. Gerald Holton and William Blanpied (eds.), *Science and Its Public: The Changing Relationship.* Boston Studies in the Philosophy of Science, Volume XXXIII. 1976.
97. Myles Brand and Douglas Walton (eds.), *Action Theory.* 1976.
98. Paul Gochet, *Outline of a Nominalist Theory of Proposition. An Essay in the Theory of Meaning.* 1980.
99. R. S. Cohen, P. K. Feyerabend, and M. W. Wartofsky (eds.), *Essays in Memory of Imre Lakatos.* Boston Studies in the Philosophy of Science, Volume XXXIX. 1976.
100. R. S. Cohen and J. J. Stachel (eds.), *Selected Papers of Léon Rosenfeld.* Boston Studies in the Philosophy of Science, Volume XXI. 1978.
101. R. S. Cohen, C. A. Hooker, A. C. Michalos, and J. W. van Evra (eds.), *PSA 1974: Proceedings of the 1974 Biennial Meeting of the Philosophy of Science Association.* Boston Studies in the Philosophy of Science, Volume XXXII. 1976.
102. Yehuda Fried and Joseph Agassi, *Paranoia: A Study in Diagnosis.* Boston Studies in the Philosophy of Science, Volume L. 1976.
103. Marian Przelecki, Klemens Szaniawski, and Ryszard Wójcicki (eds.), *Formal Methods in the Methodology of Empirical Sciences.* 1976.
104. John M. Vickers, *Belief and Probability.* 1976.
105. Kurt H. Wolff, *Surrender and Catch: Experience and Inquiry Today.* Boston Studies in the Philosophy of Science, Volume LI. 1976.
106. Karel Kosík, *Dialectics of the Concrete.* Boston Studies in the Philosophy of Science, Volume LII. 1976.
107. Nelson Goodman, *The Structure of Appearance.* (Third edition.) Boston Studies in the Philosophy of Science, Volume LIII. 1977.
108. Jerzy Giedymin (ed.), *Kazimierz Ajdukiewicz: The Scientific World-Perspective and Other Essays, 1931-1963.* 1978.
109. Robert L. Causey, *Unity of Science.* 1977.
110. Richard E. Grandy, *Advanced Logic for Applications.* 1977.
111. Robert P. McArthur, *Tense Logic.* 1976.
112. Lars Lindahl, *Position and Change. A Study in Law and Logic.* 1977.
113. Raimo Tuomela, *Dispositions.* 1978.
114 Herbert A. Simon, *Models of Discovery and Other Topics in the Methods of Science.* Boston Studies in the Philosophy of Science, Volume LIV. 1977.
115. Roger D. Rosenkrantz, *Inference, Method and Decision.* 1977.
116. Raimo Tuomela, *Human Action and Its Explanation. A Study on the Philosophical Foundations of Psychology.* 1977.
117. Morris Lazerowitz, *The Language of Philosophy. Freud and Wittgenstein.* Boston Studies in the Philosophy of Science, Volume LV. 1977.
118. Stanislaw Leśniewski, *Collected Works* (ed. by S. J. Surma, J. T. J. Srzednicki, and D. I. Barnett, with an annotated bibliography by V. Frederick Rickey). 1980. (Forthcoming.)
119. Jerzy Pelc, *Semiotics in Poland, 1894-1969.* 1978.
120. Ingmar Pörn, *Action Theory and Social Science. Some Formal Models.* 1977.
121. Joseph Margolis, *Persons and Minds. The Prospects of Nonreductive Materialism.* Boston Studies in the Philosophy of Science, Volume LVII. 1977.
122. Jaakko Hintikka, Ilkka Niiniluoto, and Esa Saarinen (eds.), *Essays on Mathematical and Philosophical Logic.* 1978.
123. Theo A. F. Kuipers, *Studies in Inductive Probability and Rational Expectation.* 1978.

124. Esa Saarinen, Risto Hilpinen, Ilkka Niiniluoto, and Merrill Provence Hintikka (eds.), *Essays in Honour of Jaakko Hintikka on the Occasion of His Fiftieth Birthday.* 1978.
125. Gerard Radnitzky and Gunnar Andersson (eds.), *Progress and Rationality in Science.* Boston Studies in the Philosophy of Science, Volume LVIII. 1978.
126. Peter Mittelstaedt, *Quantum Logic.* 1978.
127. Kenneth A. Bowen, *Model Theory for Modal Logic. Kripke Models for Modal Predicate Calculi.* 1978.
128. Howard Alexander Bursen, *Dismantling the Memory Machine. A Philosophical Investigation of Machine Theories of Memory.* 1978.
129. Marx W. Wartofsky, *Models: Representation and the Scientific Understanding.* Boston Studies in the Philosophy of Science, Volume XLVIII. 1979.
130. Don Ihde, *Technics and Praxis. A Philosophy of Technology.* Boston Studies in the Philosophy of Science, Volume XXIV. 1978.
131. Jerzy J. Wiatr (ed.), *Polish Essays in the Methodology of the Social Sciences.* Boston Studies in the Philosophy of Science, Volume XXIX. 1979.
132. Wesley C. Salmon (ed.), *Hans Reichenbach: Logical Empiricist.* 1979.
133. Peter Bieri, Rolf-P. Horstmann, and Lorenz Krüger (eds.), *Transcendental Arguments in Science. Essays in Epistemology.* 1979.
134. Mihailo Marković and Gajo Petrović (eds.), *Praxis. Yugoslav Essays in the Philosophy and Methodology of the Social Sciences.* Boston Studies in the Philosophy of Science, Volume XXXVI. 1979.
135. Ryszard Wójcicki, *Topics in the Formal Methodology of Empirical Sciences.* 1979.
136. Gerard Radnitzky and Gunnar Andersson (eds.), *The Structure and Development of Science.* Boston Studies in the Philosophy of Science, Volume LIX. 1979.
137. Judson Chambers Webb, *Mechanism, Mentalism, and Metamathematics. An Essay on Finitism.* 1980.
138. D. F. Gustafson and B. L. Tapscott (eds.), *Body, Mind, and Method. Essays in Honor of Virgil C. Aldrich.* 1979.
139. Leszek Nowak, *The Structure of Idealization. Towards a Systematic Interpretation of the Marxian Idea of Science.* 1979.
140. Chaim Perelman, *The New Rhetoric and the Humanities. Essays on Rhetoric and Its Applications.* 1979.
141. Wlodzimierz Rabinowicz, *Universalizability. A Study in Morals and Metaphysics.* 1979.

LIBRARY OF DAVIDSON COLLEGE

Books on regular loan may be checked out for **two weeks**. Books must be presented at the Circulation Desk in order to be renewed.

A fine is charged after date due.

Special books are subject to special regulations at the discretion of the library staff.